Sprinkle Irrigation of Row and Field Crops

Blaine Hanson
Irrigation and Drainage Specialist
Department of Land, Air, and Water Resources
University of California, Davis

Larry Schwankl
Irrigation Specialist
Department of Land, Air, and Water Resources
University of California, Davis

Steve Orloff
Farm Advisor, Siskiyou County
University of California Cooperative Extension

Blake Sanden
Farm Advisor, Kern County
University of California Cooperative Extension

University *of* California
Agriculture and Natural Resources
Publication 3527

Funded by the Joseph G. Prosser Trust

To order or obtain ANR publications and other products, visit the ANR Communication Services online catalog at http://anrcatalog.ucanr.edu/ or phone 1-800-994-8849. Direct inquiries to

University of California
Agriculture and Natural Resources
Communication Services
2801 Second Street
Davis, CA 95618
Telephone 1-800-994-8849
E-mail: anrcatalog@ucanr.edu

Publication 3527
ISBN-13: 978-1-60107-699-1

Illustration and photo credits given in the captions. Cover photo by Steve Orloff; design by Will Suckow.

This publication has been anonymously peer reviewed for technical accuracy by University of California scientists and other qualified professionals. This review process was managed by ANR Associate Editor for Land, Air, and Water Sciences.

POD-7/18-ED/DES/SO

Contents

Figures

Tables

Preface

This publication is one of a series of water management manuals prepared by the University of California to help California growers to address practical irrigation matters. These manuals are designed to provide in-depth information in a semitechnical format. Information on ordering these manuals can be found on the reverse side of the title page.

We thank the Joseph G. Prosser Trust for the funding to publish this manual. The endowment established in 1989 by Joseph G. Prosser and his son, Thomas W. Prosser, supports research on the efficient use of irrigation water by researchers in the University of California. In 1951, Thomas Prosser founded Irrometer, Inc. (Riverside, California), an internationally known manufacturer of soil moisture sensors.

Introduction

Sprinkle Irrigation

Sprinkle irrigation is the overhead application of water through small orifices or nozzles. Water under pressure flows through the orifice and is sprayed over the land. The spray pattern depends on the type of sprinkle irrigation system, sprinkler spacing, type of sprinkler head and nozzles, sprinkler height, sprinkler operating pressure, and wind speed and wind direction relative to the lateral line orientation of the sprinkler.

Sprinkle irrigation is used on about 29 percent of the irrigated acreage in California, primarily on vegetable crops (California Department of Water Resources, Update 2005). However, in some areas of the state, such as the intermountain area of Northern California, sprinkle irrigation is the primary irrigation method used.

Advantages

Major advantages of sprinkle irrigation include the following:

- Small applications of water are feasible.
- Frequent applications of water can occur, depending on the type of sprinkler system and design characteristics.
- Surface runoff is eliminated or reduced.
- Irrigation of steep or undulating topography is feasible.
- Irrigations can be automated for some types of sprinkle irrigation systems.
- Management or control of salts is improved.
- Some fertilizers and agricultural protection chemicals can be injected into irrigation water for some sprinkle irrigation systems.
- Sprinkle irrigation systems are easy to manage.

Disadvantages

The major disadvantages of sprinkle irrigation include the following:

- Initial capital costs are higher compared with surface irrigation costs.
- Increased energy costs can occur for pressurizing the irrigation system.
- Some types of sprinkle systems may have high labor costs to operate.
- Adverse wind effects can severely affect water application patterns.
- Foliar absorption of salts under sprinkle irrigation can reduce yield.
- Irrigating soils with very low infiltration rates may be difficult to do without surface runoff.
- Wetted canopy may enhance conditions for some diseases.

Focus of Publication

This publication provides practical information on the design, management, and maintenance of the sprinkle irrigation methods commonly used in California for irrigating field and row crops.

Hand-move, wheel-line, and portable solid-set sprinkle systems are discussed in detail because they are commonly used in California. Center-pivot, linear-move, and traveling gun systems, which are not commonly used in California, are also discussed. The publication provides information on management considerations such as when to irrigate, how much water to apply, and how to monitor soil moisture. Included are discussions on uniformity and efficiency, sprinkle lateral design considerations, calculating pressure losses along laterals, factors affecting uniformity, effect of pressure, spacing, and wind on catch-can uniformity, as well as evaluating and improving sprinkle irrigation systems. In addition, information is provided on evaporation, day versus night irrigation, pump selection, variable frequency drives, and energy cost reduction.

Types of Sprinkle Irrigation Systems

The types of sprinkle irrigation systems used in California are listed below, while characteristics are summarized in table 1.

Table 1. Summary of sprinkler system characteristics

Type	Portable?	Cost	Labor	Pressure (psi)	Uniformity	Wind effects	Soil texture*	Field shape restrictions
hand-move	yes	low	high	35–60	low to moderate	severe	M-F	none
wheel-line	yes	low to moderate	moderate	35–60	low to moderate	severe	M-F	none
solid-set	moderate	moderate	moderate	35–60	low to moderate	severe	all	none
traveling gun	yes	moderate	moderate	80–100	low to moderate	severe	M-F	none
center-pivot	no	high	low	10–60	high	low	C-M	square
linear-move	no	high	low	10–60	high	low	C-F	rectangular

Note: *M = medium-textured soil, F = fine-textured soil, C = coarse-textured soil.

Periodic-Move Sprinkle Irrigation Systems

These sprinkle systems are moved from set to set (hand-move and wheel-line systems) and from field to field (portable solid-set systems).

- *Hand-move sprinkle systems.* These systems consist of laterals of aluminum pipe with quick-release couplers, normally with end-mount risers on each section of pipe. The pipe sections are typically 30 or 40 feet long (with 30 being more common), which also is the distance between sprinklers along the lateral. The laterals are moved for each irrigation set. No overlap of sprinkler patterns occurs from an adjacent lateral during an irrigation set. Sprinkler overlap is accomplished by the placement of the lateral of the next irrigation set.
- *Wheel-line or side-roll sprinkle systems.* These systems also consist of laterals of aluminum pipe with each section supported by a large aluminum wheel. The pipe forms the axle for the wheels. The lateral is moved using an engine mounted at its center. The laterals are moved for each irrigation set.

Portable Solid-Set Sprinkle Systems

These systems use the same type of aluminum pipe as do the hand-move systems. Spacings along the mainline are such that sufficient overlap of the spray pattern occurs for each irrigation set. Enough aluminum pipe and sprinklers are available to cover the entire field. Fields are frequently split into sections or blocks, with the block or in some cases the entire field irrigated at one time. To reduce labor costs, laterals are not moved until the end of the irrigation season or after plant establishment.

Permanent Sprinkle Systems

- *Undertree systems.* These systems consist of laterals periodically buried throughout the field. Only the riser pipe and the sprinkler are visible. They are commonly used for tree crops.
- *Overhead systems.* These systems are used for frost control of vine crops.

Continuous-Move Sprinkle Systems

- *Center-pivot sprinkle systems.* These systems sprinkle water from a lateral that continuously moves in a circle and is supported by towers periodically spaced along the lateral. The towers, mounted on rubber tires, normally are driven by electric motors. Older center pivots may be driven by water or hydraulic oil motors. Some types of center-pivot systems can be moved from field to field. These systems commonly irrigate a circular area (around 125 to 130 acres) inside a quarter of a section (160 acres). A primary disadvantage of these systems is the loss of irrigated land due to the difficulty in irrigating the corners. A swing arm configuration is sometimes used by center-pivot systems to irrigate corners.
- *Linear-move sprinkle systems.* These systems use the same hardware as do center pivots except that their movement is down the length of a field instead of in a circle. An engine-driven pump is mounted on a tower, which pumps out of a supply ditch or is supplied by a large flexible hose attached to a mainline.
- *Traveling sprinkle systems, sometimes referred to as "big guns."* These systems typically use one large nozzle mounted on a movable carriage with a flow rate of up to about 1,000 gallons per minute, depending on the system design. A flexible hose supplies water to the sprinkler carriage. The carriage normally moves continuously along a travel lane or alley during an irrigation set and is manually moved from set to set. However, smaller systems that are hand-moved along the lane are sometimes used. The sprinkler carriage can be self-propelled, pulled by the hose as it is reeled up, or hand-moved in the case of the smaller systems. These systems can wet a strip up to about 400 feet wide along the length of the lane.

Sprinkle Irrigation in California

Periodic-move and portable solid-set sprinkle irrigation systems are commonly used for row and field crops in California, while permanent sprinkle systems are normally used on tree and vine crops. Continuous-move systems are not commonly used in California, although center-pivot systems are frequently used in the intermountain area of Northern California and in the high desert areas of Southern California. The limitation of center-pivot systems is that their application rate characteristics are not compatible with the water infiltration rate characteristics on many soils in the Central Valley of California. Thus, experience has shown that substantial surface runoff occurs using these systems, particularly in the sloping areas of the west side of the San Joaquin Valley. However, interest in center-pivot systems is increasing with alfalfa irrigators and in areas with level (no slope) fields in the Central Valley.

Uniformity and Efficiency of Sprinkle Irrigation Systems: General Considerations

The most important performance characteristics of sprinkle irrigation systems are uniformity and irrigation efficiency. Uniformity refers to the evenness with which water is applied or made to infiltrate throughout the field, and it depends on system design and maintenance. Irrigation efficiency refers to the ability of an irrigation system to match the applied water with that needed for crop production. Irrigation efficiency depends on uniformity and management. Higher uniformity results in a greater potential for irrigation efficiency in a properly managed irrigation system.

Uniformity

If all parts of a field received exactly the same amount of water, the uniformity would be 100 percent. However, regardless of the irrigation method, uniformities less than 100 percent occur, and thus some areas of a field receive more water than other areas. If the least-watered areas of the field receive an amount equal to the soil moisture depletion, excess amounts of water will be applied to other areas, resulting in water percolating below the root zone, commonly called deep percolation. Lower distribution uniformities result in greater differences in applied or infiltrated water throughout the field and potentially more drainage or deep percolation below the root zone.

An index commonly used to assess the uniformity of infiltrated or applied water is the distribution uniformity (DU), calculated as follows:

$$DU = \frac{100\,\overline{X}_{LQ}}{\overline{X}}$$

where:

$\overline{X}$ = the average amount of infiltrated or applied water for the entire field

$\overline{X}_{LQ}$ = the average of the lowest one-fourth of the measurements of applied or infiltrated water, commonly called the low quarter

Irrigation Efficiency

Irrigation efficiency is the ratio of the amount of water beneficially used for crop production to the amount of water applied to the field. Evapotranspiration (ET) is the largest beneficial use of irrigation water. Frost protection and leaching for salinity control are also beneficial uses.

Losses affecting the irrigation efficiency are percolation below the root zone, surface runoff, and evaporation of water as it travels through the air. Percolation occurs when the amount of infiltrated water exceeds the soil moisture storage capacity of the soil in the root zone. Surface runoff occurs when the application rate of the irrigation water exceeds the infiltration rate and may be difficult to avoid in soils with very low infiltration rates. Evaporative losses from sprinklers depend on nozzle and sprinkler design and climate characteristics such as humidity, temperature, and wind speed. Evaporative losses generally are much smaller than percolation and surface runoff losses.

The potential irrigation efficiency of a properly irrigated field is approximately equal to the DU, assuming no surface runoff.

Typical and Potential DU Values

Mobile laboratory evaluations of irrigation systems conducted throughout the state revealed the average DUs for several types of sprinkle irrigation systems used in California (table 2). Hand-move, wheel-line, and portable solid-set sprinkle systems had the lowest distribution uniformity, primarily because of wind effects on the uniformity. Continuous-move systems, such as center-pivot and linear-move, had the highest distribution uniformity. The distribution uniformity of undertree sprinklers was also high. The major source of nonuniformity for the undertree systems was sprinkler discharge variation due to pressure variation. Uniformity of applied water between sprinklers was not considered in these evaluations.

Table 2 also lists potential values of distribution uniformity (DU). The potential for hand-move systems, etc., is limited because of the wind effects on these systems. The mobile laboratory data showed few of these systems with distribution uniformities exceeding 80 percent, mainly because of the wind characteristics of many areas in California. Wind speed is the primary contributor to low uniformities. The DU of many of the continuous-move systems was greater than 80 percent. Note that the potential DU is an estimate of the potential irrigation efficiency.

Table 2. Average and potential distribution uniformities of sprinkle irrigation systems

Sprinkler system	Average DU (%)	Potential DU (%)
hand-move, wheel-line, portable solid-set	62	70–85*
undertree	79†	> 80‡
continuous-move	75	75–85

Source: Hanson 1995.

Note:

* Wind speeds smaller than about 5 miles per hour.

† Does not include the effect of the catch-can uniformity.

‡ Due to pressure losses only.

Management Considerations

Evapotranspiration, Crop Growth, and Yield

Crop water use, or evapotranspiration (ET), consists of *transpiration* and *evaporation*. Transpiration is the evaporation or loss of water from plant leaves (at least 95 percent of the water taken up by plants is lost in this manner), and evaporation is the loss of water from the soil surface. Because it is difficult to measure the two components separately, they are combined into one term.

Yield of field and row crops is directly related to ET. As ET increases, crop yield increases in a more or less direct manner with maximum yield occurring at maximum ET. Maximum ET depends on climatic characteristics. ET can be decreased by irrigation management practices that provide insufficient soil moisture, thus reducing yields of field and row crops.

Factors Affecting Evapotranspiration

Evapotranspiration (ET) depends on the following factors:

Climate

Climatic factors include solar radiation, wind speed, air humidity, and air temperature. Solar radiation is the primary source of energy for ET. An increase in solar radiation, temperature, or wind speed raises ET; an increase in air humidity decreases ET.

Plant

Plant type and stage of growth affect ET. In addition, plant health is also a factor. Healthy plants will use more water and potentially have higher yields than plants affected by nutrient deficiencies, disease, or insect damage.

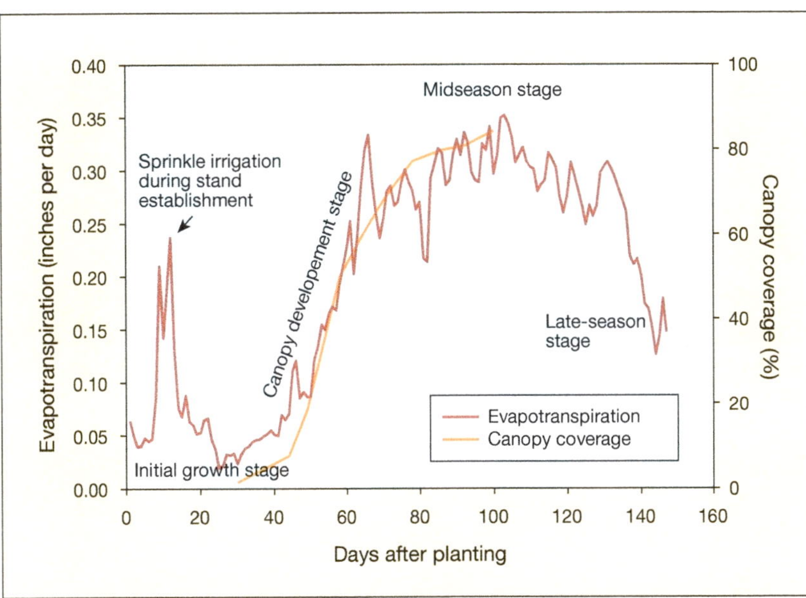

Figure 1. Evapotranspiration and canopy growth for the various growth stages of processing tomatoes.

Soil Moisture Content

Adequate soil moisture results in maximum evapotranspiration rates. Excessive soil moisture depletion due to inadequate irrigation reduces ET and thus yield.

Plant Growth and ET

During the initial growth stage of crop germination and establishment, ET consists mainly of evaporation from the soil surface (fig. 1) because little crop canopy shades the ground, thus directly exposing the soil surface to solar radiation. Sprinkle irrigation during the stand establishment growth stage causes high evaporation rates during and shortly after irrigation due to the wet soil surface (see fig. 1). Rapid canopy development follows the initial growth stage, where the canopy cover increases from about 10 percent to between 70 and 80 percent. The canopy cover is defined as the percentage of soil surface area shaded by the canopy at midday. During this period, ET also increases rapidly (see fig. 1). Adequate irrigation during this growth stage is necessary to develop a full canopy cover, resulting in maximum yields for most crops. As the canopy develops, evaporation from the soil surface decreases and transpiration from leaves increases. Maximum canopy coverage occurs during the midseason growth stage, where nearly all of the ET is transpiration and is maximum (see fig. 1). Canopy coverage typically peaks at about 80 to 90 percent for many crops. As most crops approach maturity, their canopy cover and rate of ET may decline due to aging leaves and senescence.

Estimating Crop Evapotranspiration

Daily ET is commonly calculated by

$$ET = K_C \times ET_O$$

where:
ET = the crop evapotranspiration in inches per day
K_C = the crop coefficient
ET_O = the ET of a reference crop in inches per day

Crop coefficients depend on crop type, stage of growth, and type of reference crop used (grass or alfalfa). Maximum crop coefficients of most field and row crops range from 1.0 to 1.2. Crop coefficients for some row and field crops commonly irrigated with sprinkle irrigation systems are shown in appendix B, "Crop Coefficients of Various Row Crops."

ET_o in California is the ET of an irrigated grass crop where soil moisture is not a limiting factor. This grass crop reference provides a standard measure of evapotranspiration that can be easily determined using climatic data and appropriate equations. Note that in some areas alfalfa is the reference crop (ET_r). Alfalfa ET_r values are higher than those of grass ET_o, which means that crop coefficients used with alfalfa ET_r are smaller than those used with ET_o.

The California Irrigation Management Information System (CIMIS) provides daily ET_o data, accessed at their Web site, www.cimis.water.ca.gov. However, average values of ET_o (historical ET_o), calculated using many years of measured daily values, can be used in the Central Valley without causing a significant error in ET calculations. Table A-1 (appendix A) contains average values of ET_o for a number of Central Valley locations.

When Should Irrigations Occur?

Sprinkle irrigation systems need proper irrigation scheduling or irrigation management to realize their potential. One critical management question is: "When should irrigations occur?" The answer is to irrigate when the soil moisture depletion approaches the maximum allowable.

Maximum Allowable Soil Moisture Depletion

The maximum allowable depletion is the maximum soil moisture depletion that does not reduce crop yield. It depends on crop type, soil type, and root depth. The allowable depletion is calculated by first determining the available soil moisture in the root zone, which is defined as the difference between the soil moisture at field capacity (soil moisture shortly after an irrigation) and at permanent wilting point (soil moisture at which permanent wilting of the plant occurs). Available soil moisture depends on the soil texture (table 3).

Table 3. Available soil moisture for various soil textures

Soil texture	Available soil moisture (in/ft)
sand	0.7
loamy sand	1.1
sandy loam	1.4
loam	1.8
silt loam	1.8
sandy clay loam	1.3
sandy clay	1.6
clay loam	1.6
silty clay loam	1.9
silty clay	2.4
clay	2.2

Available soil moisture frequently is expressed in inches of water per foot of soil. Multiply this value (see table 3) by the root depth to obtain the total available soil moisture in inches for the crop root zone. For soil profiles with variable soil texture, determine the available soil moisture in inches per foot for each soil texture, and multiply that value by the depth interval of each texture. Table A-2 in appendix A lists effective root depths for many crops; however, actual root depths of a particular field are influenced by site-specific conditions such as hardpans, clay lenses, shallow water tables, cultural practices, etc.

Table 4 lists maximum allowable soil moisture depletions, expressed as a percentage of the total available soil moisture, for many crops. The actual soil moisture depletion should not exceed the maximum allowable depletion, the goal being to prevent yield reductions by using an appropriate interval between irrigations.

Table 4. Approximate allowable soil moisture depletion for various crops

Crop	Allowable depletion (% of AW)	Crop	Allowable depletion (% of AW)
alfalfa	50–55	lettuce	30
barley	50–55	melon	35
bean	45	olive	65
beet	50	onion	25
broccoli	50	pasture	50–60
cabbage	45	pea	35
carrot	50	pepper	35
cauliflower	35	potato	25
celery	20	safflower	60
citrus	50	sorghum	55
clover	35	soybean	50
corn	50–60	spinach	20
corn (silage)	50–60	strawberry	15
cotton	50–65	sugar beet	50–80
cucumber	50	sunflower	45
date	50	sweet potato	65
deciduous fruit/nuts	50	tomato	40
grain		wheat	
small	50–60	vegetative	50–70
winter	50–60	ripening	90
grape	35		
grass	50		

Note: AW = available soil moisture.

Deciding When to Irrigate

A water balance is commonly used to answer this question. This approach assumes that soil moisture depletion and ET between irrigations are equal. Calculate the ET using appropriate crop coefficients and ET_0, as discussed on the previous page. One simply calculates the cumulative or total ET since the last irrigation. When the total ET after an irrigation approaches the maximum allowable soil moisture depletion, then irrigate.

Days between irrigation or irrigation intervals can be determined by dividing the maximum allowable soil moisture depletion in inches by the average daily ET in inches per day during a time period of interest, or by the following equation:

Irrigation interval = Maximum allowable soil moisture depletion ÷ daily ET

Example

Calculate the maximum allowable soil moisture depletion and irrigation interval for cotton. The soil type is a clay loam. The daily ET is 0.3 inches per day.
Calculate the allowable soil moisture depletion and irrigation interval for cotton. The soil type is a clay loam. The daily ET is 0.3 inches per day.

Step 1. Determine the available soil moisture from table 3. The available soil moisture for a clay loam is 1.6 inches per foot.

Step 2. Determine the effective root depth from table A-2 (appendix A) or from field measurements. For cotton, the root depth is 4 feet.

Step 3. Multiply the available soil moisture by the root depth to obtain the total available soil moisture. The total available soil moisture is 1.6 inches per foot × 4 feet or 6.4 inches.

Step 4. Multiply the total available soil moisture by the allowable soil moisture depletion (see table 4). The allowable soil moisture depletion for cotton is 50 percent of the total available soil moisture. The allowable depletion, then, is 6.4 inches × 50 ÷ 100 = 3.2 inches. (Note that you need to divide by 100 because the allowable depletion is expressed as a percentage.)

Step 5. Calculate the irrigation interval for an allowable depletion of 4.6 inches and a daily ET rate of 0.3 inches per day.

Irrigation interval = 3.2 inches ÷ 0.3 inches per day = 10.6 days (use 10 or 11 days)

The water balance method is inappropriate for scheduling irrigations during the initial and the early part of the canopy development stages. The allowable soil moisture criterion previously discussed is based on the root depth of a mature crop. During the early growth stages, the root depth changes rapidly over time, is smaller than that of the mature crop, and is unknown. Thus, irrigator experience and observation should be used during the early growth stages to determine when to irrigate. For example, one irrigation of tomatoes may occur during the initial growth stage while multiple irrigations may occur for lettuce and onion because of differences in root development. However, ET during the early growth stages can be estimated using the appropriate crop coefficients and ET_O, and then it can be compared to the amount of applied water to ensure sufficient water applications.

This water balance approach (i.e., irrigate when the total ET since the last irrigation equals the allowable soil moisture depletion) does not work for alfalfa irrigation. Irrigation timing of alfalfa is controlled by the harvest schedule. The first irrigation after harvest cannot occur until the hay bales are removed; the timing of the last irrigation must allow sufficient soil-drying time before the harvest. Thus, options for alfalfa irrigation between harvests are one, two, or three irrigations. The actual soil moisture depletion may exceed the maximum allowable soil moisture depletion for one irrigation between harvests; however, for two or three irrigations, actual soil moisture depletions may be smaller than the maximum allowable soil moisture depletion. Between the last irrigation before harvest and the first irrigation after harvest, the actual soil moisture depletion most likely will exceed the maximum allowable soil moisture depletion. However, the ET during a given irrigation interval can be estimated using crop coefficients and ET_0, with irrigation set times adjusted accordingly for sprinkle irrigation of alfalfa.

The water balance approach may also be inappropriate under shallow groundwater conditions, in which crop water use of the groundwater occurs. For this condition, the soil moisture depletion between irrigations may be smaller than the total ET between irrigations.

Monitoring Soil Moisture

Monitoring soil moisture can supplement the water balance method to help determine when to irrigate. It can be particularly helpful under shallow groundwater conditions and can also help to evaluate adequacy of irrigation from germination through the rapid phase of canopy development when the root depth is changing. The section "Monitoring Soil Moisture," beginning on the next page, discusses using soil moisture instruments for irrigation water management.

How Much Water Should Be Applied?

A second important question is: "How much water should be applied during an irrigation?" The answer to this question can prevent applying too little or too much water. Two methods can be used to answer this question: one is measuring the flow rate of water into the field, and the other is using the sprinkler discharge rates of sprinklers.

Field Flow Rate Method

This method is appropriate for all sprinkle irrigation systems. Data required are the flow rate of water into the field (usually measured at the pumping plant), the acres irrigated, and the irrigation time required to irrigate the field. The depth of applied water is calculated from

$$D = (Q \times T) \div (449 \times A)$$

where:
D = the average depth of applied water in inches
Q = the field flow rate in gallons per minute
T = the actual irrigation time (not including down time for moves) to irrigate the field in hours
A = acres irrigated
Flow rates measured in cubic feet per second (cfs) must be converted to gallons per minute by multiplying cfs by 449.

A flow meter is required for this method. The most common flow meter used in sprinkle irrigation systems is a propeller flow meter.

Example

Calculate the depth of applied water for a 160-acre field. The flow rate of the pumping plant is 1,200 gallons per minute. Ten days are required to irrigate the field with the pump operating 22 hours per day. The total operating time of the pumping plant is 10 days × 22 hours per day or 220 hours per field irrigation.

$$D = (1{,}200 \text{ gpm} \times 220 \text{ hours}) \div (449 \times 160 \text{ acres}) = 3.7 \text{ inches}$$

Sprinkler Discharge Rate Method

The amount of applied water by a sprinkle irrigation system can be calculated using

$$D = AR \times T$$

where:

D = the depth of applied water in inches of water

AR = the application rate of the sprinkle irrigation system in inches per hour

T = the irrigation set time in hours

The application rate of periodic-move sprinkle irrigation systems depends on the discharge rate of a single sprinkler and the sprinkler spacing. Calculating the application rate is discussed in detail in the section "Determining Application Rates Using Sprinkler Discharge Rates."

Monitoring Soil Moisture

Soil moisture monitoring can help evaluate the adequacy of current irrigation practices and is especially valuable when used in conjunction with ET_0-based estimates of ET. It provides an assessment of the actual conditions in the soil and can be used to adjust irrigation scheduling based on ET_0-based ET calculations. It is particularly useful where accurate site-specific weather data and crop coefficients used to estimate ET do not exist.

Soil Moisture Sensors

Soil moisture can be monitored by sampling the soil or using sensors such as tensiometers, electrical resistance blocks, dielectric sensors (for which there are many types), and neutron moisture meters. This section will focus on tensiometers and electrical resistance sensors because of their relatively low cost and ease of use, which make them particularly appropriate for irrigator use. Detailed information on soil moisture sensors can be found in *Monitoring Soil Moisture for Irrigation Water Management,* ANR Publication 21635 (Hanson et al. 2007). This publication can be ordered at the ANR Catalog Web site, anrcatalog.ucdavis.edu.

Tensiometers may be the best choice for crops sensitive to small soil moisture depletions, such as lettuce, onions, celery, carrots, and potato. Tensiometers measure soil moisture tension, which is the tenacity at which soil moisture is retained by the soil. (Note that the drier the soil, the higher the soil moisture tension is.) Tensiometers are easily installed and read, but they can be damaged and require more maintenance than other sensors.

Electrical resistance sensors determine soil moisture tension by measuring the electrical resistance of the water in a porous material, commonly called a block. Several types of electrical resistance sensors are available. Gypsum sensors or blocks are composed of gypsum only; Watermark sensors (Irrometer, Inc. Riverside, CA) consist of a porous material (sand and ceramic) containing a small wafer of gypsum. The sensors are buried in the ground.

Electrodes embedded in the block material measure the electrical resistance. Calibration equations are used to convert electrical resistance to soil moisture tension. Wire leads connect the electrodes in the buried sensor to a portable meter or data logger, which measures the electrical resistance between the electrodes and, in some cases, converts the readings to soil moisture tension. The electrical resistance depends on the water content of the sensor. The higher the water content, the smaller the electrical resistance is. The sensors take up and release moisture as the soil wets and dries, which changes their electrical resistance. As the soil dries, water flows out, thus increasing the electrical resistance. The readings of the portable instrument or data logger are correlated with soil moisture tension values using calibration data provided by the manufacturer.

Resistance sensors may not be useful for very coarse sandy soils because water flow in and out may be too slow to accurately reflect the rapid changes that occur in soil moisture tension in these soils. In these cases, a tensiometer designed for responding quickly in sandy soils is recommended. Also, some gypsum blocks may not respond to changes in the soil moisture until dry soil conditions occur.

The electrical resistance also depends on the salinity of water. The higher the salinity, the smaller the resistance is. Gypsum in the block dissolves as water flows into the sensor, thus buffering the effect of soil water salinity on the sensor's readings by maintaining a constant salinity in the sensor's water. However, saline irrigation water may affect the readings by increasing the salinity of the water flowing into the sensor.

Watermark sensors are particularly appropriate for agricultural fields. These sensors are easy to install and read, require no maintenance, and respond reasonably well to changes in soil moisture of wet soils. Both the portable instrument and stationary data loggers used with these sensors have been calibrated to provide soil moisture tension values in centibars (a measurement of tension). The lower the centibar value, the higher the soil moisture content will be; and conversely, the higher the centibar value, the smaller the moisture content of the soil will be.

How to Install Watermark Sensors

Proper installation of the sensors is important for valid readings. Good contact with the soil is required so that water moves readily between the sensor and the soil. To achieve proper installation, the following procedure is recommended:

Step 1. Soak the sensor in water for a few minutes or the time recommended by the manufacturer. Then check the sensor reading to ensure that it is operating correctly. The reading should indicate that the sensor is saturated (a reading near zero for the Watermark sensor).

Step 2. Use a soil probe (with a diameter only slightly larger than that of the sensor) to make a hole slightly deeper than the desired depth for the sensor.

Step 3. Mix soil and water (and a small amount of gypsum, if possible, to help buffer any salinity effects on the readings) to make a slurry. Pour about a quarter of a cup of the slurry into the hole. The slurry is necessary to create good contact between the soil and the block.

Step 4. Push the sensor into the slurry at the bottom of the hole using a ½-inch diameter PVC pipe (Schedule 40 or 80). Run the wire through the pipe to prevent damage to the wire lead and keep the wire taut as the sensor is pushed down the hole to prevent the wire from falling down into the hole. If, instead, the wire is run along the side of the pipe, cut a notch in the bottom of the pipe and position the wire in the notch so it is not damaged when the sensor is being pushed into the soil.

Step 5. Remove the pipe and backfill the hole with the soil that was removed, taking care not to damage the wire leads. Using the PVC pipe, tamp down the soil periodically while backfilling. Use a mechanism such as knots in the wire lead or a tag to identify the depth of the sensor. Wait at least 24 hours after installation before taking readings to allow the sensor to equilibrate with the soil moisture.

An alternative installation of the Watermark sensor involves gluing the sensor to a ½-inch Class 315 PVC pipe. This gauge pipe properly fits the end of the sensor. An epoxy cement is preferred, but clear PVC cement works as well. Take care that the small hole in the plastic at the top of the block is not plugged, because this hole allows air to flow in and out of the block as water moves in and out of the block. A plugged hole will greatly reduce the response time of the block to changes in soil moisture. The advantage of this type of installation is that it allows for easy removal and reuse of the sensors.

Sensor Location, Number, and Depth

At least two sites of sensors per 40 acres are recommended. More sites may be needed, depending on the soil texture variability in a field. The greater the variability, the greater the number of sensor locations needed. Select areas of the field with uniform crop growth that are most representative of the entire field. It is also important to select areas that receive good sprinkler coverage.

The sensors must be placed at the proper depth to represent the soil moisture status in the crop root zone. At minimum, install one sensor at a depth of about one-fourth to one-third of the root zone for irrigation scheduling, and install another sensor near the bottom of the root zone to assess the depth of irrigation. For relatively deep-rooted crops such as tomatoes, alfalfa, or cotton, installing sensors at equal depth intervals down to the bottom of the root zone is preferable for a more accurate assessment of the depth of wetting and soil moisture extraction patterns. Install the sensors not more than 6 inches from the plant row for row crops.

Using Soil Moisture Tension Readings

Soil moisture tension readings can be useful in several ways. They can determine the soil moisture status at the start of the growing season; indicate the depth at which irrigation water infiltrates during irrigation; assist in the decision to irrigate; and assess the adequacy of current irrigation practices.

Assessing soil moisture status at the start of the growing season. Knowing the soil moisture status at the beginning of the growing season is necessary for determining when to start irrigating, even if an ET_0-based technique is used for irrigation scheduling. The ET_0-based approach assumes a full profile of soil moisture at the start of the crop season. However, assessing the soil moisture status can be difficult without a tool like electrical resistance sensors or tensiometers. The sensors are useful for determining whether the soil profile is full after winter or spring precipitation and when to begin irrigating.

Depth of water infiltration. Installing sensors at different depths helps assess the depth of infiltration during an irrigation. The soil moisture tension will decrease as the infiltrating water reaches an installation depth. Sensors placed at shallower depths will respond first, followed by sensors at the deeper depths as infiltration progresses downward. A decrease in soil moisture tension near the bottom of the root zone during an irrigation indicates that the soil profile is full.

Deciding when to irrigate. Soil moisture tension readings indicate when irrigation is needed. Irrigation should occur as the soil moisture tension approaches a specified value (table 5). When sensors are installed at different depths, use the uppermost sensor to schedule irrigations, because that is typically where most roots are located and it is the zone of greatest water extraction. The soil moisture tension value used to trigger an irrigation should account for the soil type as well as the crop. Irrigate at a lower soil moisture tension in a sandier soil than in a soil

Table 5. Recommended maximum soil moisture tension for various crops

Crop	Tension (centibars)	Crop	Tension (centibars)
alfalfa	80–150	grain (small)	
broccoli		vegetative	40–50
early	45–55	ripening	70–80
postbud	60–70	grapes	
cabbage	60–70	early	40–50
cantaloupe	35–40	mature	100
carrot	55–65	lettuce	40–60
cauliflower	60–70	onion	45–65
celery	20–30	potato	30–50
citrus	50–70	strawberry	20–30
corn (sweet)	50–80	tomato	60–150
deciduous tree	50–80		

Note: Irrigate as the soil moisture tension approaches the value for a specific crop. Use the larger values in cool, humid conditions and the smaller values in warm, dry conditions.

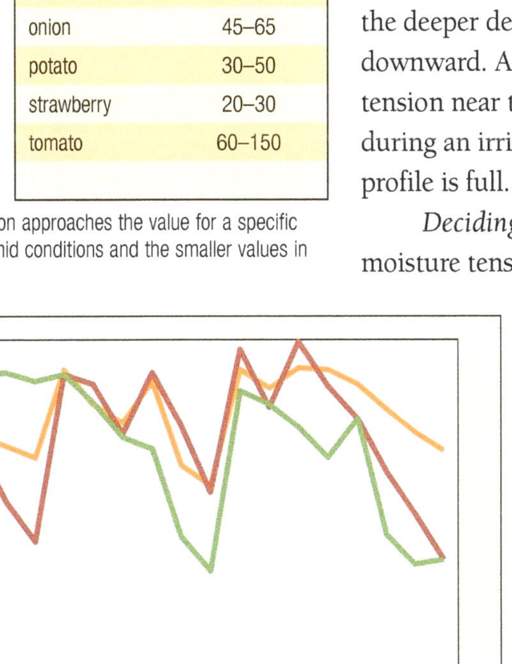

Figure 2. An example of proper irrigation scheduling. Irrigations generally occurred when the soil moisture tension reached the shaded area, which represents the recommended values for alfalfa.

with a greater water-holding capacity. Grower experience may be needed to determine the most appropriate value at which to irrigate for different soil textures.

Assessing the adequacy of current irrigation practices. Data on soil moisture tension plotted over time can present a picture of the soil moisture status throughout the irrigation season. A spreadsheet with a graphics capability is the easiest way to view the data. Ideally, soil moisture tension should approach but not exceed the maximum recommended value (see table 5) and should be near zero just after an irrigation, particularly at the shallower depths (fig. 2). If soil moisture tensions never reach the recommended value, irrigations are too frequent and the field is over-irrigated (fig. 3). This indicates that the interval between irrigations could be lengthened and/or less water applied per irrigation. Conversely, if soil moisture tension values exceed the recommended value, too little water is being applied, thus causing under-irrigation, and/or the intervals between irrigations are too long (fig. 4).

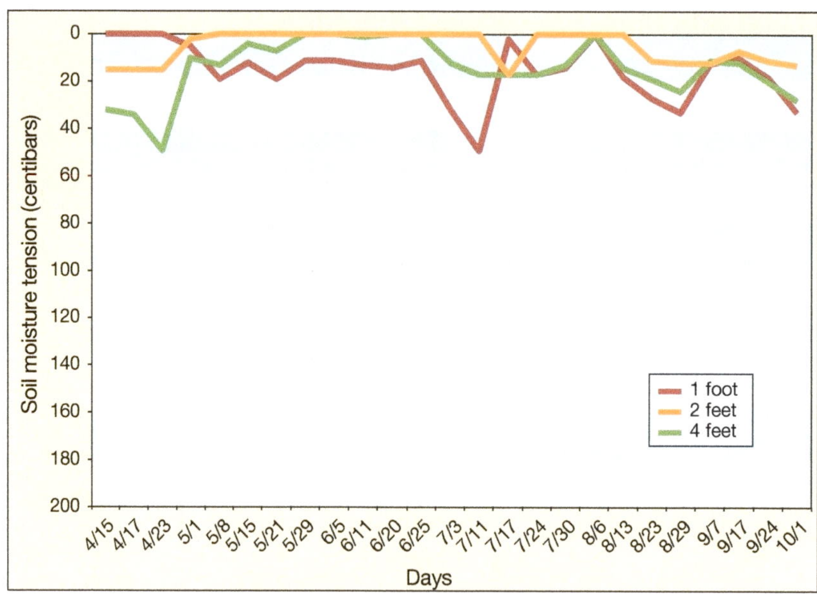

Figure 3. An example of over-irrigation. Irrigations occurred at soil moisture tensions much smaller than the recommended values for alfalfa.

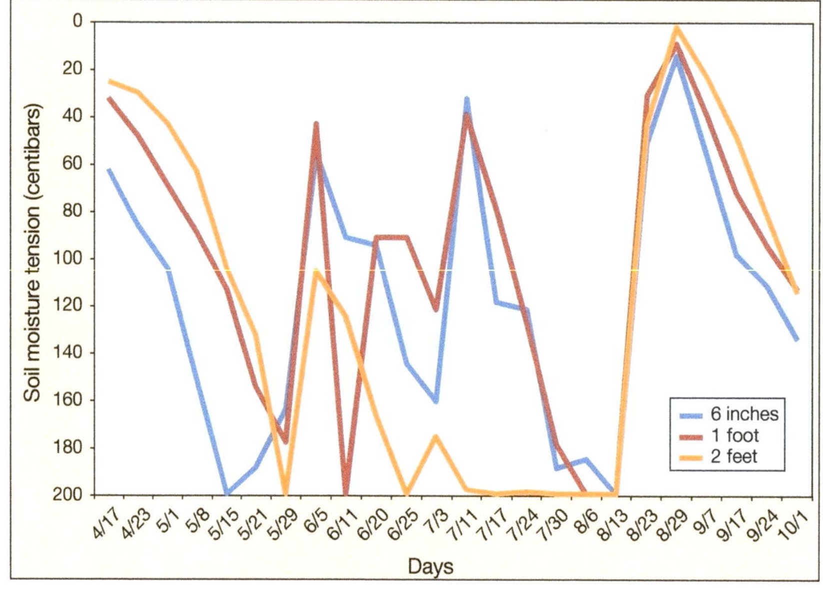

Figure 4. An example of under-irrigation. Irrigations occurred when soil moisture tension values greatly exceeded the recommended values for alfalfa.

Soil moisture monitoring has some limitations for irrigation scheduling. Soil moisture readings are site specific and reflect a particular location in the field, which may not be representative of the entire field due to variability in soil type, crop growth, and irrigation water applications. Also, soil moisture monitoring cannot be used for estimating the time of future irrigation events, as can be done with the water balance approach using historical ET_0. It indicates only the current moisture status, not what it will be in a few days.

Periodic-Move and Portable Solid-Set Sprinkle Irrigation Systems

Periodic-Move Sprinkle Irrigation Systems

Periodic-move sprinkle irrigation systems are commonly used in California to irrigate row and field crops. They are classified as hand-move systems, wheel-line or side-roll systems, and portable solid-set sprinkle systems. The portable solid-set systems are normally used for irrigating high cash value crops, while the others are used for crops such as cotton, alfalfa, and pasture.

Hand-Move Sprinkle Irrigation

Hand-move sprinkle irrigation systems (fig. 5) involve dividing the field into blocks, with one sprinkler lateral per block. All blocks are normally irrigated at the same time, but each irrigation set only irrigates a fraction of the block. At the end of an irrigation set, the sprinkler laterals are picked up by hand and moved to a new set. The distance that the sprinkler lateral is moved equals the sprinkler spacing along the mainline or submain. Aluminum pipe, ranging from 2 to 4 inches in diameter and in sections 20 to 40 feet long, is frequently used for the sprinkler laterals. Quick-release couplers are installed on each section of pipe to facilitate moving the lines. The mainlines may be buried or aboveground, while submains and laterals are aboveground. At the end of the irrigation season, all the aboveground aluminum pipe is removed from the field.

Figure 5. Hand-move sprinkle irrigation system. *Photo:* Blaine Hanson.

After the last irrigation set, the sprinkler laterals must be returned back to the initial starting point of each block prior to the next field irrigation. This rotation prevents over-irrigating the most recently irrigated areas of the field.

Sprinkler spacings of 30 feet (spacing along the lateral) by 40 feet (spacing along the mainline or submain) or 30 by 45 feet are normally used for cotton and vegetable crops, while spacings of 40 by 60 feet are frequently used for alfalfa and pasture. Small sprinkler nozzle sizes of $^{7}/_{64}$, $^{1}/_{8}$, and $^{9}/_{64}$ inch in diameter are normally used for the smaller spacings. Larger nozzles sizes of 3⁄16 to 13⁄64 inch are commonly used for the larger spacings.

Wheel-Line or Side-Roll Sprinkle Irrigation Systems

Wheel-line sprinkle irrigation systems are commonly used for alfalfa irrigation and for stand establishment of tomatoes in the Sacramento Valley. They consist of aluminum lateral pipes rigidly coupled together and mounted on large aluminum wheels (fig. 6A). A flexible hose connects the laterals to the mainline or submain. The lateral pipe is the axle of the system, with the wheel spacing equal to the sprinkler spacing, and the sprinklers located midway between wheels. The sprinkler lateral is moved with an engine mounted at the center of the line, which rotates the pipe and causes the lateral to roll sideways, hence the common name of side-roll sprinklers. Wheel-line systems are best suited for fields that are rectangular with relatively uniform topography.

The move distance depends on the wheel diameter and the number of wheel revolutions. Lateral moves of 60 feet are common, which requires four revolutions of a 4.8-foot diameter wheel (circumference equals 15 feet). Before moving the lateral, the pipe must be drained using quick drains, installed at each sprinkler location, that open automatically when the pressure drops inside the pipeline.

The layout of wheel-line sprinkle irrigation systems is similar to that of hand-move systems, with the field divided into blocks and one sprinkler lateral per block. At the end of each irrigation set, the sprinkler lateral is rolled to the location of the next irrigation set. After the last irrigation of a block, the sprinkler lateral must be rolled back to the initial starting point of each block prior to the next field irrigation to prevent over-irrigating the most recently irrigated parts of the field.

Figure 6. Wheel-line sprinkle irrigation system on alfalfa (A), and wheel-line sprinkle irrigation system with trailer lines on tomatoes (B). *Photos:* Steve Orloff (A); Blaine Hanson (B).

Wheel-line laterals frequently are about 1,300 feet long. A 4-inch diameter pipe is commonly used; however, 5-inch pipe is also used, which results in less pressure loss along the lateral length. A sprinkler spacing of 40 feet is normally used along the lateral, with a 60-foot lateral spacing along the mainline. However, sometimes a 30-foot sprinkler spacing and a 50-foot lateral spacing are used.

Some irrigators modify their wheel-line systems by attaching trailer lines to the lateral pipe to increase the irrigated area per irrigation set (fig. 6B). The trailer line, attached at each sprinkler location along the lateral, extends perpendicular from the lateral and may contain one or more sprinklers. The trailer line is pulled by the rotating lateral to the new irrigation set. This approach is used by some tomato irrigators to increase the area irrigated during an irrigation set and reduce field irrigation time of newly planted tomatoes.

Sprinkler nozzles normally used for wheel-line systems are 5/32, 11/64, 3/16, and 13/64 inch. In some cases, a small nozzle, called a spreader nozzle, is also used along with the larger nozzle. Self-levelers are recommended for wheel-line systems to ensure that the sprinkler remains upright after a move.

Portable Solid-Set Sprinkle Irrigation Systems

Portable solid-set sprinkle irrigation systems are normally used for vegetable crops (fig. 7). They may be used for stand establishment only or for the entire irrigation season. Sufficient pipe and sprinklers are used so that the entire field is covered by the irrigation system. The field may be divided into blocks, but an entire block is irrigated at the same time. Sprinkler laterals are not moved between irrigation sets. The sprinkler system is removed after stand establishment or after the irrigation season, depending on its use.

Figure 7. Portable solid-set sprinkle irrigation system. *Photo:* Larry Schwankl.

Generally, this type of sprinkle irrigation system uses sprinkler spacings ranging from 30 to 45 feet. Relatively small sprinkler nozzle sizes (such as 3/32, 1/8, 7/64, and 9/64 inch) are also used. Aluminum pipe is used for the above-ground laterals and submains. Depending on the design of the irrigation system, mainlines may be buried.

Aluminum pipe is commonly used for portable solid-set systems. However, the use of plastic pipe is increasing. The plastic pipe reduces the leakage at the joints that normally occurs with aluminum pipe, particularly after an irrigation has ended.

General Design Considerations

The design of the sprinkle irrigation system affects the field-wide uniformity of applied water and the application rate. Once a sprinkler system is installed, correcting any design problems can be difficult and expensive. Thus, the best approach is to install a properly designed system from the beginning.

Uniformity

The DU (distribution uniformity) of a properly designed sprinkle system should be about 75 to 85 percent under low wind conditions. This design uniformity depends on sprinkler spacings, pressure, and selection of sprinkler head and nozzles.

System Flow Rate

The system flow rate should supply the desired amount of water needed for irrigation during the period of maximum crop water use or ET. This flow rate depends on crop type, crop growth rate, acres irrigated, hours desired to irrigate the field, and the climate conditions.

Design Application Rate

The design application rate depends on the discharge rate of the sprinkler nozzle(s) and the spacings of the sprinklers. Design application rates should be used that do not cause excessive surface runoff while meeting the peak ET needs of the crop.

Pressure

Adequate sprinkler pressure should occur throughout the irrigation system. Insufficient or excessive pressure causes poor breakup of the water jet leaving the sprinkler nozzle.

Pressure Losses and Pipeline Design

Pressure will vary throughout the irrigation system due to friction losses in the pipelines and elevation differences. This variation affects pressure, application rate, and uniformity of the applied water. Appropriate diameters and lengths of mainlines, submains, and laterals should be selected to prevent excessive pressure losses due to friction losses caused by water flowing in the pipelines. These pressure losses are particularly sensitive to pipeline diameter. Proper design, which depends on pipe diameter and length, flow rate, and sprinkler spacing, is particularly important for good irrigation uniformity. Pressure losses along laterals should not exceed about 20 percent of the design pressure.

Sprinkler Spacings

The sprinkler spacing affects application rate and field-wide uniformity. The most commonly used sprinkler spacings reflect a compromise between system cost and uniformity of applied water determined from experience. Thus, only a few spacing configurations are generally used, which provide good design uniformity at a reasonable cost. Smaller spacings may only slightly improve the uniformity, but they can greatly increase the system and management costs. Larger spacings can reduce system costs, but they may reduce yield due to poor uniformity.

Pump Selection

Select a pump that will provide the desired flow rate and total head at maximum efficiency. This requires information on the pump performance characteristics and the well behavior. Pump performance characteristics can be obtained from pump dealers and installers. The behavior of the well can be determined through a pump test, which provides information on pumping levels and discharge rates.

Uniformity of Periodic-Move Sprinkle Systems

Uniformity refers to how evenly the irrigation system applies water throughout the field. The level of uniformity achievable depends on the design of the sprinkle irrigation system.

Factors Affecting Uniformity

The main factors affecting uniformity of periodic-move sprinkle irrigation systems are

- pressure
- sprinkler spacings
- wind speed and direction
- sprinkler type and nozzle type
- sprinkler rotation rate
- crop interference
- malfunctioning sprinkler heads or worn nozzles
- nonvertical or leaning risers
- irrigation set time

The uniformity of the water distribution pattern between sprinklers primarily reflects the operating pressure, sprinkler spacings, and wind speed and direction, although crop interference and nonvertical risers can have a significant effect. However, the uniformity of applied water is dominated by high wind speeds, regardless of the other factors. All of these factors interact to contribute to the pattern, and thus the distribution of applied water must be determined by a catch-can test. This involves placing containers between sprinklers, measuring the amount of water collected in the containers during an irrigation, and evaluating the water distribution between sprinklers from these catch-can data.

Single Irrigation versus Seasonal Catch-Can Uniformity

The uniformity between sprinklers normally is characterized by a one-time catch-can test made during an irrigation. The results reflect the wind speed and direction, pressure, etc., at the time of the test. A seasonal uniformity value, obtained by making multiple catch-can tests throughout the irrigation season, will differ from the single event measurement mainly because of variation of pressure, wind speed, and wind direction throughout the irrigation season. Research on carrots irrigated with a portable solid-set sprinkle irrigation system showed that the uniformity based on single-irrigation events underestimated the seasonal uniformity by 5 to 15 percent. Single event DUs ranged from 71.8 to 86.0 percent, while seasonal DUs ranged from 81.6 to 90.8 percent.

Effect of Uniformity on Crop Yield

In theory, nonuniform water applications should result in nonuniform soil moisture and in turn nonuniform crop yield, with the smallest yield and soil moisture occurring for the smallest amount of applied water. In reality, this is not the case for situations where the average amount of applied water in the least-watered part of the field at least equals the crop evapotranspiration.

Little correlation was found between catch-can uniformity and uniformity of carrot yield for a portable solid-set sprinkle system. Higher yield uniformity occurred compared to applied water uniformity, indicating that more of the crop received adequate water than one might think based on the applied water uniformity values. A study of sprinkle-irrigated alfalfa showed moderate correlation between applied water uniformity and soil moisture uniformity. However, better correlation occurred between soil moisture and yield than between applied water and yield. Alfalfa yield uniformity values were about 95 percent, whereas applied water values ranged from about 66 to 77 percent. Similar behavior occurred with corn, where the uniformities of yield and soil moisture were higher than that of the applied water. One possible reason for this behavior is that the amount of water applied to the least-watered area was at least equal to the soil moisture depletion, and thus, root-zone soil moisture values throughout the test areas would be similar regardless of the distribution of applied water. A redistribution of soil moisture area after the irrigation is another possible explanation for the higher uniformities of yield and soil moisture.

Since yield uniformity greatly exceeded applied water uniformity in these studies, is the uniformity of the applied water important? The answer is "Yes." Applying sufficient water so that the least-watered areas of the field receive an amount equal to the soil moisture depletion means that the rest of the field receives too much water, which causes deep percolation and excessive leaching of nutrients. The higher the uniformity of applied water, the smaller will be the average amount of applied water needed to adequately irrigate the field, which in turn will reduce both deep percolation and irrigation costs.

On the other hand, applying the same amount of water regardless of the uniformity will result in more of the field being deficit-irrigated for small values of DU compared to high values. This could reduce field-wide yields for crops sensitive to deficit irrigation.

Pressure

Pressure affects the sprinkler discharge rate and the pattern or distribution of water between sprinklers.

Sprinkler Discharge Rate

The sprinkler discharge rate depends on the sprinkler pressure and the diameter of the nozzle(s). The relationship between discharge rate, pressure, and nozzle size is

$$q = K \times \sqrt{P}$$

where:
q = the sprinkler discharge rate in gallons per minute (gpm)
P = the pressure in pounds per square inch (psi)
K = a sprinkler coefficient that depends on the nozzle diameter and units used for pressure and discharge rate

Effect on Distribution of Applied Water

Sprinkler pressure and nozzle size affect the pattern of the applied water and the wetted distance. Water leaves the sprinkler nozzle as a jet of water. Because of turbulent eddies in the jet, the jet surface begins to break up. After the jet is sufficiently disrupted, the action of air resistance further breaks up the jet into water droplets. Water near the edge of the jet produces smaller droplets, while water near the core of the jet produces the largest droplets.

Water pressure plays a major role in the droplet size distribution. Figure 8 shows cross-sections of profiles of applied water for high, low, and satisfactory pressures. High pressure produces smaller droplets that fall near the sprinkler and result in large amounts of water near the sprinkler (fig. 8A). Low pressure results in larger water droplets that travel further from the sprinkler and tend to fall in a ridge of applied water some distance from the sprinkler (fig. 8B). This is commonly called a donut-shaped pattern. A satisfactory pressure causes a triangular-shaped pattern (fig. 8C).

Nozzle size of the sprinkler also affects the distribution of droplets, but the effect is less than that of pressure. For a given pressure, larger nozzles cause larger droplets. However, a small nozzle operating at low pressure will produce larger droplets than a larger nozzle operating at a higher pressure.

The wetted diameter of a sprinkler increases with pressure and nozzle diameter (fig. 9). Remember, however, that the pattern of applied water changes with

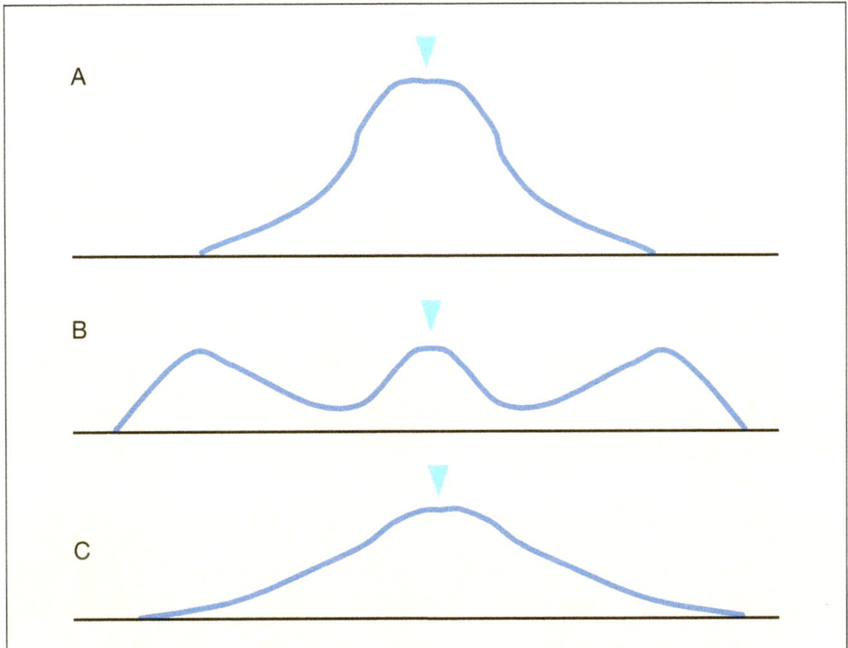

Figure 8. Cross-sections of applied water distributions for high pressure (A), low pressure (B), and adequate pressure (C). The black inverted triangle is the sprinkler head location.

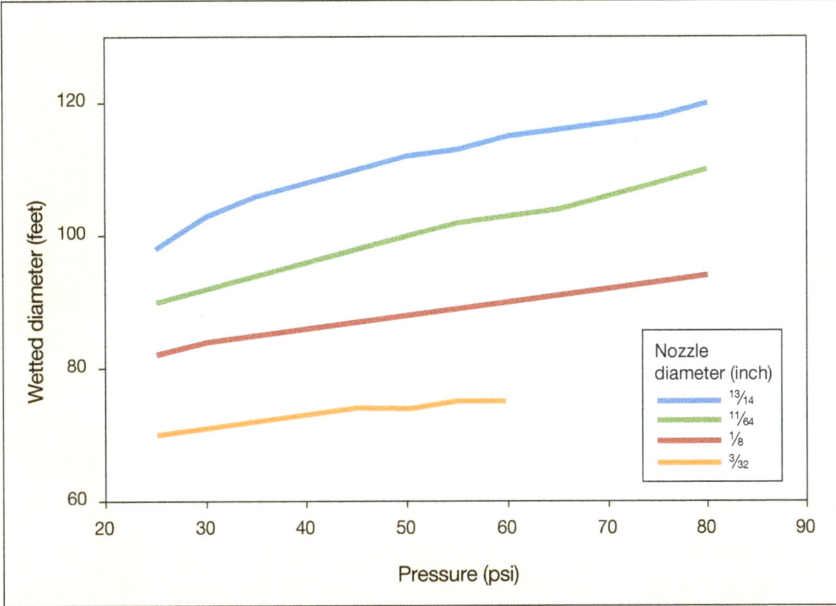

Figure 9. Wetted diameter of a single-impact sprinkler for different nozzle diameters and different pressures (psi).

pressure. Thus, for a high pressure, the wetted diameter may be relatively large, but little water may be applied in the outer area of the pattern due to small droplet sizes at higher pressures.

Recommended pressures vary with sprinkler nozzle diameter. Some recommended pressures are listed in table 6. Manufacturers of sprinklers should be contacted about the performance of their specific sprinklers.

Table 6. Recommended sprinkler pressures for standard circular nozzles

Nozzle diameter (in)	Pressure range (psi)
3⁄32	30–40
1⁄8	30–50
5⁄32	30–55
11⁄64	30–55
3⁄16	35–60
13⁄64	35–60

Determining the Flow Rate

The desired flow rate of the sprinkle irrigation system depends on the crop evapotranspiration rate, the number of acres being irrigated, the efficiency of the irrigation system, and the hours needed to irrigate the field. The desired flow rate can be estimated by the following equation:

$$Q = (449 \times A \times D_m) \div T$$

where:

Q = the pump flow rate in gallons per minute

A = the acres irrigated

D_m = the desired depth of water in inches

T = the hours needed to irrigate the field

449 = a conversion factor used for these units

The desired depth of application (D_m) should meet the peak ET during the crop season and also account for the irrigation efficiency. This value can be estimated using historical CIMIS ET_o (table A-1 in appendix A) and appropriate crop coefficients in appendix B. Note that the time during the year when the peak ET occurs will depend on the crop, planting time, and climate. For example, the peak ET of lettuce will differ from that of tomatoes in the San Joaquin Valley. However, a system flow rate determined using summer peak ET values of crops with maximum crop coefficients of 1.1 or greater will be sufficient for meeting the peak ETs during nonsummer periods.

The time required to irrigate the field is a management decision based on the interval between irrigations. The interval can be calculated using the procedure in the section "When Should Irrigations Occur?" Irrigating continuously during the period of peak ET will result in the smallest design flow rate.

A procedure for calculating the desired flow rate is as follows:

Step 1. Determine the historical ET_o and crop coefficient during the period of peak ET. Historical ET_o is found in table A-1 (in appendix A), and crop coefficients are in appendix B.

Step 2. Determine the daily peak ET.

Step 3. Determine the total ET between irrigations during the period of peak ET.

Step 4. Determine the desired depth of application (D_m) accounting for the irrigation efficiency (IE) where D_m = Total $ET \div IE$. Use an irrigation efficiency equal to 75 percent for periodic-move sprinkle irrigation systems.

Step 5. Calculate the desired flow rate using the above equation.

Example

Determine the flow rate needed to irrigate 160 acres of cotton. The plan is to irrigate the field every 10 days for 22 hours per day for a total of 220 hours for the field irrigation. The location is near Five Points, along the west side of the San Joaquin Valley.

Step 1. Determine the ET_o and crop coefficient for the period of peak ET. From figure B-4 (in appendix B), select the crop coefficient of 1.1 for cotton. From table A-1 (in appendix A), select an ET_o equal to 0.30 inches per day for the period of peak ET, which occurs during the latter part of June or the earlier part of July at Five Points.

Step 2. Calculate the peak daily ET. ET = 1.1 × 0.30 inches per day = 0.33 inches per day.

Step 3. Determine the total ET between irrigations.
ET total = 10 days × 0.33 inches per day = 3.3 inches.

Step 4. Determine D_m using an irrigation efficiency equal to 75 percent.
D_m = 3.3 ÷ 0.75 = 4.4 inches.

Step 5. Calculate the desired flow rate.
Q = (449 × 160 acres × 4.4 inches) ÷ 220 hours = 1,437 gallons per minute.

Selecting Sprinkler Heads and Nozzles

Factors affecting the selection of sprinkler heads and nozzles include desired discharge rate, water distribution patterns, crop considerations, wind conditions, and pressure.

Sprinkler Head

The most common type of sprinkler head used for sprinkle irrigation systems of row and field crops is the impact sprinkler. These sprinkler heads are durable and reliable, and they can provide good uniformity of applied water for properly designed irrigation systems. The durability is important because of the rough handling conditions that result from moving the sprinkler laterals from set to set and to other fields.

Impact sprinklers rotate in a stop/start sequence, which is due to the movement of an impact spoon that uses the energy of the water jet for rotation. They normally rotate 360 degrees, but sprinkler heads are available for a part-circle rotation.

Other types of sprinkler heads include gear-driven heads that rotate continuously (normally used for landscape irrigation systems) and heads that consist of a slowly rotating deflector plate with slots that produce streams of water. The deflector plate may have a single slot or multiple slots. These sprinkler heads are commonly used for center-pivot and linear-move sprinkle systems.

Nozzle Arrangement and Size

The size of the nozzle(s) should provide the discharge needed for the design application rate. A variety of nozzle sizes are available for impact sprinklers, thus providing flexibility in the irrigation system design. Nozzle sizes range from 1⁄64 to 9⁄32 inch, depending on the manufacturer. Smaller nozzle sizes are used for vegetable crops, while larger nozzle sizes are used for alfalfa and other forage crops.

Some impact sprinkler heads have two nozzles: a range nozzle and a spreader or spray nozzle. The range nozzle, the larger of the nozzles, applies water along the outer part of a wetted area. The spreader nozzle applies water closer to the sprinkler. Larger water drop sizes usually occur for range nozzles, which increases the wetted diameter. Smaller droplet sizes occur for spreader nozzles, which provides a better water distribution near the sprinkler. One objective for using the two nozzles is to improve the uniformity of applied water, although little information is available on uniformity differences between single-nozzle and double-nozzle sprinkler heads. Another purpose is to provide larger sprinkler discharge rates without using nozzles with very large diameters.

Some sprinkler heads contain straightening vanes placed just upstream of the nozzle, which can increase the wetted diameter and uniformity of application.

Nozzle Shape

A nozzle with a circular orifice shape with a constant cross-sectional area along its bore length is commonly used for impact sprinklers. Nozzles with noncircular orifice shapes (triangular, square, rectangular, or circular with notches around the orifice edge) are designed to provide better uniformity of applied water under low-pressure conditions. Field research, however, suggests that these nozzles may not be very effective in improving the uniformity of applied water under low-pressure conditions.

Flow-Control Nozzles

Flow-control nozzles contain a flexible orifice that changes diameter as pressure changes, and they are discussed in further detail in the section "Flow-Control Nozzles."

Trajectory

The trajectory of the jet of water should be considered for a particular set of operating conditions. A variety of trajectories are available, ranging between 0 and 28 degrees. The appropriate trajectory depends mainly on crop and wind. The smaller the trajectory angle, the smaller the wetted diameter. Low trajectories are recommended under consistently windy conditions. Under varying wind conditions, an intermediate trajectory angle is suggested. Under consistently low wind conditions, high trajectory angles can be used, which will result in a larger wetted diameter than that of lower-trajectory sprinkler heads.

Determining Application Rates Using Sprinkler Discharge Rates

The average application rate depends on the sprinkler discharge rate (which also depends on the pressure, nozzle size, and number of nozzles on a sprinkler head) and the sprinkler spacings. It is calculated by

Table 7. Sprinkler application rate for different nozzle sizes, spacings, and pressures

Pressure (psi)	7/64	7/64	1/8	9/64	3/32	5/32	11/64	3/16	13/64
	(30×30)	(30×40)	(30×40)	(30×40)	(30×50)	(40×60)			
	Application rate (in/hr)								
30	0.202	0.151	0.198	0.254	0.089	0.153	0.184	0.218	0.264
35	0.219	0.164	0.215	0.272	0.096	0.165	0.198	0.236	0.285
40	0.235	0.183	0.230	0.292	0.103	0.176	0.213	0.253	0.305
45	0.248	0.186	0.245	0.309	0.109	0.188	0.225	0.264	0.325
50	0.262	0.197	0.258	0.321	0.115	0.197	0.233	0.281	0.341
55	0.277	0.207	0.272	0.341	0.121	0.208	0.249	0.293	0.357
60	0.289	0.217	0.284	0.355	0.126	0.216	0.261	0.313	0.373

Note: column headers are Nozzle size (in); second header row is Sprinkler spacing (ft).

$$AR = (96.3 \times q) \div (S_l \times S_m)$$

where:

AR = the application rate in inches per hour

q = the sprinkler discharge rate in gallons per minute

S_l = the sprinkler spacing along the lateral in feet

S_m = the lateral spacing along the mainline in feet

96.3 = a conversion constant for specific units used for spacing and discharge rate

The amount of applied water is calculated by

$$D = AR \times T$$

where:

D = the inches of applied water

T = the irrigation set time in hours

The sprinkler discharge rate (q) can be obtained from sprinkler manufacturers' catalogues or measured directly by using the method described in the section "Evaluating Sprinkle Irrigation System Performance."

The sprinkler application rate can also be determined from table 7 for a range of commonly used nozzle sizes and sprinkler spacings. Information required is the sprinkler spacings and the sprinkler nozzle pressure and diameter. For spacings not listed in table 7, use the above equation and the nozzle discharge rates in table 8 or measured discharge rates.

Table 8. Sprinkler nozzle discharge rates for various nozzle sizes and pressures

Nozzle size (in)	30	35	40	45	50	55	60	65
	Discharge rate (gpm)							
3/32	1.7	1.5	1.6	1.7	1.8	1.9	2.0	2.1
7/64	1.9	2.0	2.2	2.3	2.4	2.6	2.7	2.8
1/8	2.5	2.7	2.9	3.1	3.2	3.4	3.6	3.7
9/64	3.1	3.4	3.6	3.8	4.1	4.2	4.4	4.6
5/32	3.9	4.2	4.5	4.7	5.0	5.2	5.5	5.8
11/64	4.7	5.1	5.4	5.7	6.1	6.3	6.6	6.9
3/16	5.6	6.0	6.4	6.8	7.2	7.6	7.9	8.2
13/64	6.5	7.1	7.6	8.1	8.5	9.2	9.2	9.5

Note: column headers are Pressure (psi).

Where should the sprinkler discharge rate be measured? Measuring the sprinkler discharge rates where the least amount of water is applied in the field (area where the pressure is the lowest) ensures that most of the field is adequately irrigated.

Example

Calculate the application rate and the depth of water applied by a sprinkle irrigation system with a sprinkler discharge rate of 6 gallons per minute, a sprinkler spacing along the lateral of 40 feet, and a lateral spacing along the mainline of 60 feet. The irrigation set time is 12 hours.
The application rate is

$$AR = (96.3 \times 6 \text{ gpm}) \div (40 \text{ feet} \times 60 \text{ feet}) = 0.24 \text{ inches per hour,}$$

and thus the amount of water applied in 12 hours is

$$D = 0.24 \text{ inches per hour} \times 12 \text{ hours} = 2.9 \text{ inches}$$

Design Application Rate

The application rate of a sprinkler system is relatively constant during an irrigation set. However, the infiltration rate of water into the soil water is highest early in the irrigation and declines to smaller rates during the irrigation. Over time, the infiltration rate becomes relatively constant and is commonly called the basic infiltration rate.

Surface runoff occurs when the application rate of a sprinkle irrigation system exceeds the infiltration rate. Thus, the design application rate should be based on the infiltration rate that occurs near the latter part of the irrigation set. Some recommended design application rates are in table 9.

Once a design application rate is selected, the sprinkler discharge needed to achieve that application rate can be calculated from

$$q = (AR_d \times S_l \times S_m) \div 96.3$$

where:
AR_d = the design application rate

Table 9. Recommended application rates for sprinklers

	Application rate (in/hr)	
Soil texture	**0–5% slope**	**5–8% slope**
coarse sandy	2.00	1.50
sandy loam	1.00	0.80
silt loam	0.50	0.40
clay loam	0.15	0.10

Source: USDA 1983.

For soil types not listed in table 9, the recommended application rate can be estimated using the information in the table as a guide. For example, the infiltration rate of a silty clay loam soil would be between that of a silt loam and a clay loam. Based on these infiltration rates, an application rate for a silty clay loam might be about 0.3 inches per hour.

Pressure Losses in Sprinkler Irrigation Laterals

Elevation differences and friction losses in the pipeline due to flowing water cause pressure losses in a sprinkle irrigation system. Friction losses depend on the flow rate in the pipeline, pipe diameter, roughness of the interior of the pipe, and pipeline length. Elevation differences create a 1 psi pressure change for every 2.31 feet of changed elevation. Pressure changes in sprinkler lines can change both the average application rate and the uniformity of the applied water.

Examples of Measured Pressure along Sprinkler Laterals

Most of the pressure change measured along a 1,300-foot sprinkler lateral occurred over the first half of the length (fig. 10). Thereafter, little change in pressure occurred. A progressively decreasing lateral flow rate due to sprinkler discharge caused this behavior, which is normal behavior for a properly designed lateral. Little elevation change occurred over this lateral length.

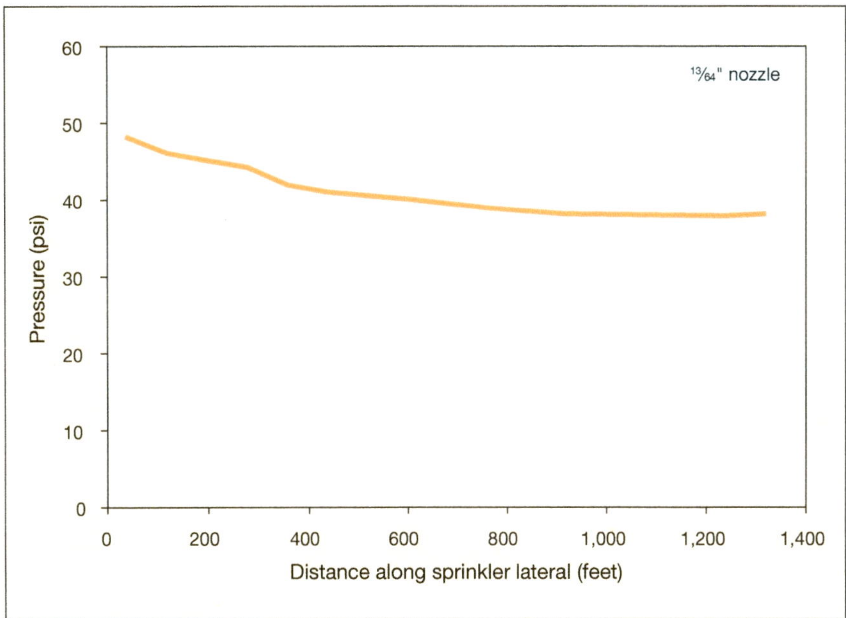

Figure 10. Pressure (psi) along a 1,300-foot sprinkler lateral. Pipe diameter was 4 inches.

Water flowing uphill reduced pressure due to both elevation differences and friction losses (fig. 11). About one half of the total pressure loss along the lateral was caused by the elevation difference.

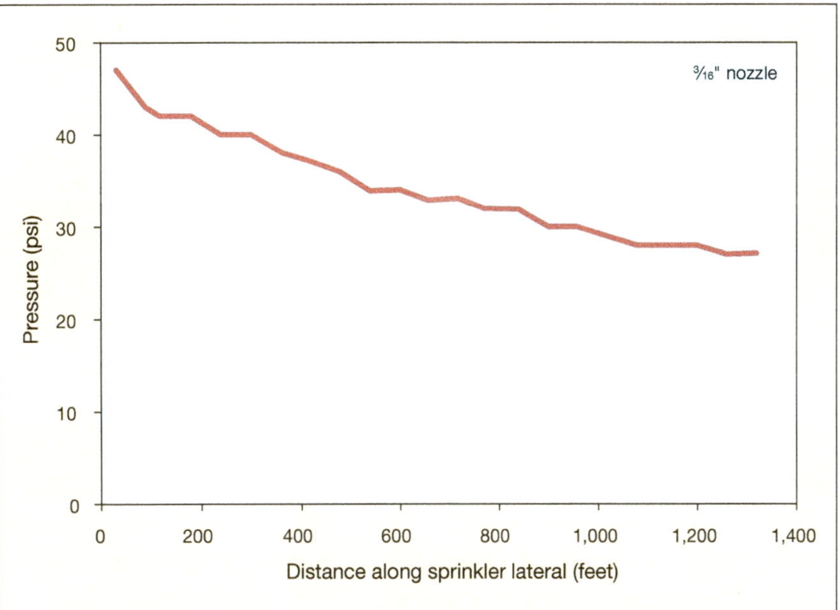

Figure 11. Pressure (psi) along a 1,300-foot lateral length with water flowing uphill. Pressure decreased due to increasing elevation along the lateral and to friction losses. About one half of the total pressure loss was caused by the elevation difference.

Near the end of a 2,600-foot-long lateral, sprinkler pressure increased because the effect of decreasing elevation on pressure was greater than the pressure loss due to friction near the end of the lateral (fig. 12).

Reducing Pressure Losses

The best approach for minimizing pressure losses is to install a properly designed sprinkle irrigation system using appropriate pipeline lengths and pipeline diameters for the desired flow rates. This involves knowing the pressure losses due to friction and the elevation differences throughout the field. Correcting problems caused by a poor design after a system is installed may be difficult and expensive.

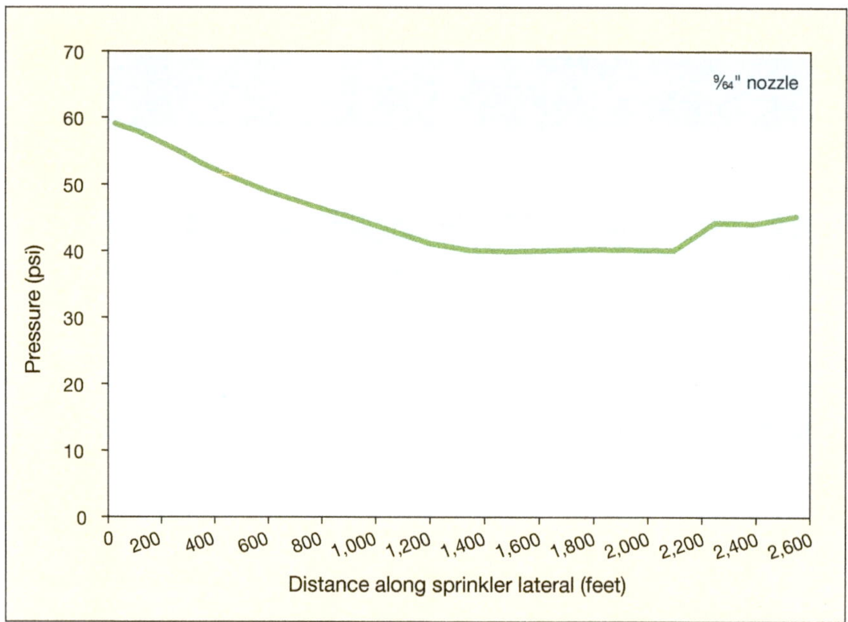

Figure 12. Pressure (psi) along a 2,600-foot lateral with water flowing downhill. Pressure decreased until about 1,400 feet due to friction losses along the lateral. Along the lower part of the lateral, pressure increased due to a net gain in pressure. This was because friction losses were smaller than the pressure increase due to the elevation difference.

Calculating Pressure Changes Due to Elevation Differences

Elevation differences affect the sprinkler pressure in sprinkle irrigation systems. A decrease in elevation between two locations along a pipeline causes a gain in pressure. An increase in elevation between two points causes a loss in pressure.

Determining Elevation Differences

The first step is to determine the elevation difference in feet along the pipeline length. Devices that can be used to determine elevation differences include GPS units, surveying levels and rods, and hand-held levels and measuring sticks.

Calculating Pressure Changes Due to Elevation Differences

The pressure change (P_e) in pounds per square inch (psi) due to an elevation difference is calculated from

$$P_e = \text{elevation difference} \div 2.31$$

where:

2.31 = the height of the column water that results in a 1 psi pressure at its base

Calculating Head and Pressure Losses Due to Friction Losses

Flowing water in a pipeline causes friction losses, and this reduces pressure. The extent of these losses depends on water flow rate, pipe diameter, roughness of the pipe wall, and length of the pipeline. Proper design of laterals and mainlines or submains of sprinkle irrigation systems requires that these pressure losses be determined.

Calculating Head Loss Due to Friction

Pressure or head losses due to friction can be calculated by the Hazen-Williams equation or the Scoby equation. These equations are

$$\text{Hazen-Williams: } H_t = \{[1050 \times (Q \div C)^{1.852}] \times (L \div 100)\} \div D^{4.87}$$

$$\text{Scoby: } H_t = [(0.287 \times K_s \text{ x } Q^{1.90}) \times (L \div 100)] \div D^{4.90}$$

where:
H_t = the head loss in feet along the pipeline length
L = the pipeline length in feet
Q = the flow rate in gallons per minute
D = the pipe internal diameter (ID) in inches
C = the pipe wall roughness coefficient for the Hazen-Williams equation (table 10)
K_S = the roughness coefficient for the Scoby equation (table 11)

The constants "1050" and "0.287" in the respective equations are conversion factors used for units of feet (head, pipeline length), gallons per minute (flow rate), and inches (pipe diameter). Different conversion factors should be used for units other than those listed above. A calculator with an exponent key is required to use these equations. The Hazen-Williams equation is more commonly used, but the Scoby equation is included because it is slightly easier to use, and roughness coefficients for aluminum pipe with couplers are available for pipe sections of 20, 30, and 40 feet. Note that there are other equations used for calculating head loss in pipelines, but they are more complicated to use.

Table 10. Coefficients for the Hazen-Williams equation

Pipe material	Coefficient
plastic	
4 inches in diameter or greater	150
2 and 3 inches	140
< 2 inches	135
cement asbestos	140
aluminum (30-foot sections with couplers)	130
galvanized steel	
new	130
old	100

Table 11. Coefficients for the Scoby equation

Pipe material	K_s
aluminum (no couplers)	0.33
aluminum (20-foot section with couplers)	0.43
aluminum (30-foot section with couplers)	0.40
aluminum (40-foot section with couplers)	0.39
steel (4-,5-, and 6-inch diameter)	0.32

Note: K_S = the roughness coefficient for the Scoby equation.

Friction Losses—Aluminum Pipe

An alternative to using the equations for calculating head losses due to friction in aluminum pipe is to use tables 12, 13, and 14, which show the head losses (feet of head loss per 100 feet of pipeline length) in aluminum pipe for section lengths of 20, 30, and 40 feet and pipe diameters of 3, 4, 5, 6, 8, 10, and

Table 12. Friction loss (feet per 100 feet) in aluminum pipe (20-foot sections, with couplers), determined from the Scoby equation

3- and 4-inch pipes

	Nominal pipe diameter, inches (actual interior diameter, inches)	
Flow rate (gpm)	3 (2.914)	4 (3.906)
50	1.11	0.26
60	1.56	0.37
70	2.10	0.50
80	2.70	0.64
90	3.38	0.80
100	4.13	0.98
110	4.95	1.18
120	5.84	1.39
130	6.80	1.62
140	7.83	1.86
150	8.92	2.12
160	10.09	2.40
170	11.32	2.69
180	12.62	3.00
190	13.98	3.33
200	15.41	3.67
210	16.91	4.02
220	18.47	4.40
230	20.10	4.78
240	21.79	5.19
250	23.55	5.60
260	25.37	6.04
270	27.26	6.49
280	29.21	6.95
290	31.22	7.43
300	33.30	7.92
310		8.43
320		8.96
330		9.50
340		10.05
350		10.62
360		11.20
370		11.80
380		12.42
390		13.04
400		13.69
410		14.34
420		15.02
430		15.70
440		16.40

5- to 12-inch pipes

	Nominal pipe diameter, inches (actual interior diameter, inches)				
Flow rate (gpm)	5 (4.90)	6 (5.885)	8 (7.856)	10 (9.818)	12 (11.818)
200	1.21	0.49	0.12	0.04	0.02
250	1.85	0.75	0.18	0.06	0.02
300	2.61	1.06	0.26	0.09	0.03
350	3.50	1.43	0.35	0.12	0.05
400	4.51	1.84	0.45	0.15	0.06
450	5.64	2.30	0.56	0.19	0.08
500	6.89	2.81	0.68	0.23	0.09
550	8.25	3.37	0.82	0.27	0.11
600	9.74	3.97	0.96	0.32	0.13
650	11.34	4.62	1.12	0.38	0.15
700	13.05	5.32	1.29	0.43	0.17
750	14.88	6.07	1.47	0.49	0.20
800	16.82	6.86	1.66	0.56	0.23
850	18.87	7.70	1.87	0.63	0.25
900	21.04	8.58	2.08	0.70	0.28
950	23.31	9.51	2.31	0.77	0.31
1,000	25.70	10.48	2.54	0.85	0.34
1,050	28.20	11.50	2.79	0.94	0.38
1,100	30.80	12.56	3.05	1.02	0.41
1,200		14.82	3.60	1.21	0.49
1,300		17.26	4.19	1.40	0.57
1,400		19.87	4.82	1.62	0.65
1,500		22.65	5.50	1.84	0.74
1,600		25.60	6.21	2.08	0.84
1,700		28.73	6.97	2.34	0.94
1,800			7.77	2.61	1.05
1,900			8.61	2.89	1.16
2,000			9.49	3.18	1.28
2,100			10.41	3.49	1.41
2,200			11.38	3.82	1.54
2,400			13.42	4.50	1.81
2,600			15.63	5.24	2.11
2,800			17.99	6.03	2.43
3,000			20.51	6.88	2.77
3,200			23.18	7.78	3.13
3,400				8.73	3.52
3,600				9.73	3.92
3,800				10.78	4.35
4,000				11.88	4.79
4,200				13.04	5.26
4,400				14.24	5.74
4,600				15.50	6.25
4,800				16.80	6.77
5,000				18.16	7.32

Table 13. Friction loss (feet per 100 feet) in aluminum pipe (30-foot sections, with couplers), determined from the Scoby equation

3- and 4-inch pipes

	Nominal pipe diameter, inches (actual interior diameter, inches)	
Flow rate (gpm)	3 (2.914)	4 (3.906)
50	1.03	0.24
60	1.46	0.35
70	1.95	0.46
80	2.51	0.60
90	3.14	0.75
100	3.84	0.91
110	4.60	1.10
120	5.43	1.29
130	6.32	1.50
140	7.28	1.73
150	8.30	1.97
160	9.38	2.23
170	10.53	2.51
180	11.74	2.79
190	13.01	3.09
200	14.34	3.41
210	15.73	3.74
220	17.18	4.09
230	18.70	4.45
240	20.27	4.82
250	21.91	5.21
260	23.60	5.62
270	25.36	6.03
280	27.17	6.47
290	29.04	6.91
300	30.98	7.37
310		7.84
320		8.33
330		8.83
340		9.35
350		9.88
360		10.42
370		10.98
380		11.55
390		12.13
400		12.73
410		13.34
420		13.97
430		14.61
440		15.26

5- to 12-inch pipes

	Nominal pipe diameter, inches (actual interior diameter, inches)				
Flow rate (gpm)	5 (4.90)	6 (5.885)	8 (7.856)	10 (9.818)	12 (11.818)
200	1.12	0.46	0.11	0.04	0.02
250	1.72	0.70	0.17	0.06	0.02
300	2.43	0.99	0.24	0.08	0.03
350	3.25	1.33	0.32	0.11	0.04
400	4.19	1.71	0.41	0.14	0.06
450	5.24	2.14	0.52	0.17	0.07
500	6.41	2.61	0.63	0.21	0.09
550	7.68	3.13	0.76	0.25	0.10
600	9.06	3.69	0.90	0.30	0.12
650	10.55	4.30	1.04	0.35	0.14
700	12.14	4.95	1.20	0.40	0.16
750	13.84	5.65	1.37	0.46	0.19
800	15.65	6.38	1.55	0.52	0.21
850	17.56	7.16	1.74	0.58	0.23
900	19.57	7.98	1.94	0.65	0.26
950	21.69	8.85	2.15	0.72	0.29
1,000	23.91	9.75	2.37	0.79	0.32
1,050	26.23	10.70	2.60	0.87	0.35
1,100	28.65	11.69	2.84	0.95	0.38
1,200		13.79	3.35	1.12	0.45
1,300		16.05	3.89	1.31	0.53
1,400		18.48	4.48	1.50	0.61
1,500		21.07	5.11	1.71	0.69
1,600		23.82	5.78	1.94	0.78
1,700		26.73	6.48	2.17	0.88
1,800			7.23	2.42	0.98
1,900			8.01	2.62	1.08
2,000			8.83	2.96	1.19
2,100			9.69	3.25	1.31
2,200			10.58	3.55	1.43
2,400			12.49	4.19	1.69
2,600			14.54	4.88	1.97
2,800			16.73	5.61	2.26
3,000			19.08	6.40	2.58
3,200			21.57	7.23	2.92
3,400				8.12	3.27
3,600				9.05	3.65
3,800				10.03	4.04
4,000				11.05	4.46
4,200				12.13	4.89
4,400				13.25	5.34
4,600				14.41	5.81
4,800				15.63	6.30
5,000				16.89	6.81

Table 14. Friction loss (feet per 100 feet) in aluminum pipe (40-foot sections, with couplers), determined from the Scoby equation

3- and 4-inch pipes

Flow rate (gpm)	Nominal pipe diameter, inches (actual interior diameter, inches) 3 (2.914)	4 (3.906)
50	1.00	0.24
60	1.42	0.34
70	1.90	0.45
80	2.45	0.58
90	3.07	0.73
100	3.75	0.89
110	4.49	1.07
120	5.30	1.26
130	6.17	1.47
140	7.10	1.69
150	8.09	1.93
160	9.15	2.18
170	10.26	2.44
180	11.44	2.72
190	12.68	3.02
200	13.98	3.33
210	15.34	3.65
220	16.75	3.99
230	18.23	4.34
240	19.76	4.70
250	21.36	5.08
260	23.01	5.48
270	24.72	5.88
280	26.49	6.30
290	28.32	6.74
300	30.20	7.19
310		7.65
320		8.12
330		8.61
340		9.12
350		9.63
360		10.16
370		10.70
380		11.26
390		11.83
400		12.41
410		13.01
420		13.62
430		14.24
440		14.88

5- to 12-inch pipes

Flow rate (gpm)	Nominal pipe diameter, inches (actual interior diameter, inches) 5 (4.90)	6 (5.885)	8 (7.856)	10 (9.818)	12 (11.818)
200	1.10	0.45	0.11	0.04	0.01
250	1.67	0.68	0.17	0.06	0.02
300	2.37	0.97	0.23	0.08	0.03
350	3.17	1.29	0.31	0.11	0.04
400	4.09	1.67	0.40	0.14	0.05
450	5.11	2.09	0.51	0.17	0.07
500	6.25	2.55	0.62	0.21	0.08
550	7.49	3.05	0.74	0.25	0.10
600	8.83	3.60	0.87	0.29	0.12
650	10.28	4.19	1.02	0.34	0.14
700	11.84	4.83	1.17	0.39	0.16
750	13.49	5.50	1.34	0.45	0.18
800	15.25	6.22	1.51	0.51	0.20
850	17.12	6.98	1.69	0.57	0.23
900	19.08	7.78	1.89	0.63	0.26
950	21.14	8.63	2.09	0.70	0.28
1,000	23.31	9.51	2.31	0.77	0.31
1,050	25.57	10.43	2.53	0.85	0.34
1,100	27.94	11.40	2.76	0.93	0.37
1,200		13.44	3.26	1.09	0.44
1,300		15.65	3.80	1.27	0.51
1,400		18.02	4.37	1.47	0.59
1,500		20.54	4.98	1.67	0.67
1,600		23.22	5.63	1.89	0.76
1,700		26.06	6.32	2.12	0.85
1,800			7.05	2.36	0.95
1,900			7.81	2.62	1.06
2,000			8.61	2.89	1.16
2,100			9.45	3.17	1.28
2,200			10.32	3.46	1.40
2,400			12.17	4.08	1.65
2,600			14.17	4.75	1.92
2,800			16.32	5.47	2.21
3,000			18.60	6.24	2.52
3,200			21.03	7.05	2.84
3,400				7.91	3.19
3,600				8.82	3.56
3,800				9.78	3.94
4,000				10.78	4.34
4,200				11.82	4.77
4,400				12.92	5.21
4,600				14.05	5.67
4,800				15.24	6.14
5,000				16.47	6.64

12 inches. These values were calculated with the Scoby equation. These friction losses include the couplers for connecting sections. The tables also include the ID (internal diameter) for the various pipe diameters (in parentheses). The total head loss (loss over the entire length of pipeline), using the values in the tables, is calculated by

$$H_t = H_{100} \times L \div 100$$

where:

H_{100} = the head loss per 100 feet found in the tables

Friction Losses—Plastic Pipe

Plastic pipe is characterized by its pressure rating or class, SDR (standard dimension rating), or schedule. The class of a pipe, e.g., Class 100, Class 125, Class 160, is the pressure rating of the pipe. A Class 125 pipe has a pressure rating of 125 pounds per square inch (psi). Plastic pipe of the same class or SDR rating has the same pressure rating regardless of pipe diameter. The inside diameter decreases and the wall thickness increases as the class increases for a nominal pipe size. For a given nominal pipe diameter, the outside diameter (OD) is the same for all class ratings. This allows the same-sized pipe fittings to be used on all classes of pipe.

The SDR rating corresponds with the class rating. The SDR is the ratio of the internal diameter to the wall thickness. It decreases as the class increases.

The schedule rating is used for plastic pipe that has the same outside diameter and wall thickness as steel pipe for the same nominal pipe size. The pressure rating varies with pipe size for the schedule-rated pipe.

Plastic pipe is also characterized by IPS (iron pipe size) and PIP (plastic irrigation pipe) designations. IPS plastic pipe has the same outside diameter (OD) as steel or iron pipe for a particular nominal rating, while PIP plastic pipe has a slightly smaller OD.

Table 15. Internal pipe diameters (inches) for IPS and PIP plastic pipe for various diameters and classes of pipe

	Nominal pipe diameter (inches)						
	3	4	5	6	8	10	12
IPS							
class 100	—	4.280	5.291	6.300	8.200	10.226	12.128
class 125	3.284	4.224	5.221	6.217	8.005	10.088	11.966
class 160	3.230	4.154	5.133	6.115	7.961	9.924	11.720
class 200	3.166	4.072	5.033	5.993	7.805	9.728	11.538
schedule 40	3.068	4.026	5.047	6.065	7.981	10.200	11.814
schedule 80	2.900	3.826	4.814	5.761	7.625	9.564	11.376
PIP							
SDR81 (80 psi)	—	—	—	5.898	7.840	9.800	11.760
SDR41 (100 psi)	—	—	—	5.840	7.762	9.702	11.642
SDR32.5 (125 psi)	—	—	—	5.762	7.658	9.572	11.486

Sources: Hunter Industries, Inc. 2009 (IPS); Diamond Plastics Corporation 2005 (PIP).

Because of the variety of methods of characterizing plastic pipe, tables of friction losses are not provided in this publication. Friction losses will need to be calculated using the Hazen-Williams equation for desired pipe size and pipe classification. Table 15 summarizes the nominal pipe IDs for different pipe classifications and sizes to help facilitate calculating friction losses.

Convert to Pressure Loss

The values in tables 12, 13, and 14, or those calculated with the Hazen-Williams or Scoby equations, are feet of head loss. The head in feet is converted to pressure using the following equation:

$$P_t = H_t \div 2.31$$

This equation is based on the relationship in which a column of water 2.31 feet high creates a pressure of 1 psi at its base.

Multiple Outlet Factor

Smaller pressure losses occur in pipelines with multiple outlets, such as sprinkler laterals. Water discharged through sprinklers along the lateral length progressively reduces the flow rate in the lateral, thus reducing pressure losses due to friction. A multiple outlet factor adjusts the pressure loss, calculated with the above equations or tables, to account for the multiple outlets. This factor depends on the number of outlets or sprinklers along a lateral and the distance of the first sprinkler from the mainline or submain. Table 16 lists multiple outlet factors recommended for sprinkle systems. The head loss or pressure loss for multiple outlet pipelines is calculated by

$$H_a = F \times H_t \text{ or } P_a = F\,P_t$$

where:

F = the multiple outlet factor

H_a, P_a = the adjusted head or pressure loss

Table 16. Multiple outlet factors for pipelines

Number of outlets	S_1	$S_{½}$
1	1.00	1.00
2	0.64	0.52
3	0.53	0.44
4	0.49	0.41
5	0.46	0.40
6	0.44	0.39
7	0.43	0.38
8	0.42	0.38
9	0.41	0.37
10–11	0.40	0.37
12–14	0.39	0.37
15–20	0.38	0.36
21–35	0.37	0.36
> 35	0.36	0.36

Note: These values depend not only on the number of sprinklers along the lateral, but also on the distance between the first sprinkler and the beginning of the lateral. Use the S_1 column if this distance equals the sprinkler spacing along the lateral; use the $S_{½}$ column if this distance equals one-half of the sprinkler spacing.

Example

Calculate the pressure loss along an aluminum sprinkler lateral with 33 sprinklers. The lateral length is 1,320 feet, lateral inflow rate is 200 gpm, and the lateral diameter is 4 inches (ID = 3.906 inches). The first sprinkler is 20 feet from the lateral inlet. The sprinkler spacing is 40 feet.

Head losses calculated by the Scoby equation

From table 11

$$K_s = 0.39 \text{ for a 40-foot length of aluminum pipe } H_t$$

$$= [(0.287 \times K_s \times Q^{1.90}) \times (L \div 100)] \div D^{4.90}$$

$$= [(0.287 \times 0.39 \times 200^{1.90}) \times (1{,}320 \div 100) \div 3.906^{4.90} = 2{,}636 \times 13.2 \div 793 = 43.9 \text{ feet}$$

Head loss calculated using table 14

From table 14, the head loss per 100 feet is 3.33 feet for a 4-inch diameter with a flow rate of 200 gallons per minute. The total loss (not adjusted for multiple outlets) is

$$H_t = 3.33 \text{ feet} \times 1{,}320 \text{ feet} \div 100 = 43.9 \text{ feet}$$

Pressure loss

The pressure loss is

$$P_t = 43.9 \text{ feet} \div 2.31 = 19.0 \text{ psi}$$

Adjusted pressure loss due to multiple outlets

From table 16, the adjustment factor for 33 outlets (number of sprinklers) is 0.36. The adjusted head loss and pressure loss is

$$H_a = 0.36 \times 43.9 \text{ feet} = 15.8 \text{ feet or } P_a = 0.36 \times 19.0 \text{ psi} = 6.8 \text{ psi}$$

Sprinkle Irrigation System Lateral Design

The lateral design should provide sufficient pressure and good uniformity of applied water along its length. Design considerations are lateral length, pipe diameter, design pressure, spacing of sprinklers, and lateral flow rate.

Allowable Pressure Loss

A common recommendation is to design sprinkler laterals such that the pressure loss along the lateral does not exceed 20 percent of the design pressure.

Calculating Pressure Losses

Procedures for determining these losses are in the previous sections "Calculating Head and Pressure Losses Due to Friction Losses" and "Calculating Pressure Changes Due to Elevation Differences."

Design Pressure

The required design inlet pressure of a lateral is the pressure needed to obtain the desired application rate and can be determined using one of the following procedures:

1. Assume that the design pressure occurs at the end of the lateral. This approach is appropriate for irrigating higher cash value crops that are sensitive to inadequate soil moisture. This approach ensures that an adequate amount of water is applied along the entire lateral length. The required inlet pressure of the lateral is calculated by

   $$P_i = P_d + P_f \pm P_e$$

   where:
   P_i = the lateral inlet pressure
   P_d = the design pressure
   P_f = the pressure loss due to friction
   P_e = the pressure gain or loss due to elevation differences
   For water flowing downhill, subtract P_e. Add P_e for water flowing uphill.

2. Assume that the design pressure is the average pressure along the lateral. The average pressure generally occurs at about one-third of the lateral length from the inlet. An exception is for downward flow on steep slopes, where the average pressure could occur at about half of the

lateral length. This approach is appropriate for alfalfa, pasture, cotton, and other sprinkle-irrigated crops that can tolerate some water stress without adversely affecting yield. This approach means that under-irrigation will occur downstream from the point of average pressure. The required inlet pressure, which will be smaller than that of the first approach, is calculated by

$$P_i = P_d + 0.75 \times P_f \pm 0.5 \times P_e$$

Design Lateral Length

The shorter the length, the smaller the pressure losses along the lateral are. The design length will depend largely on the field size and shape as well as the layout of the sprinkle irrigation system. A common approach to ensure relatively short lateral lengths is to install the mainline down the middle of the field with laterals branching off both sides of the mainline.

Design Diameter

Friction losses are particularly sensitive to diameter. The larger the diameter, the smaller the pressure losses will be, and thus, too, the energy requirements for system operation. An increase of 1 inch in the pipe diameter can decrease the friction losses by at least 35 percent. However, trade-offs exist between the costs of energy and the cost of pipe. Larger diameters result in smaller energy costs, but they will increase upfront capital costs of the irrigation system and may make it difficult to move hand-move sprinkle systems.

Design Procedure

A design procedure is presented below. This procedure assumes that the design pressure occurs at the end of the lateral.

Step 1. Calculate the lateral flow rate. The lateral flow rate is

$$Q = N \times q$$

where:
Q = lateral flow rate (gallons per minute)
N = number of sprinklers on lateral
q = design sprinkler discharge rate (gallons per minute)

If the irrigation system flow rate is known, then the lateral flow rate can also be calculated by dividing the system flow rate by the number of laterals of the irrigation system. Note that this approach assumes a constant sprinkler discharge rate along the lateral, which is not true because of friction losses and elevation differences.

Step 2. Calculate the pressure loss due to friction along the lateral using the Hazen-Williams or Scoby equations or the tables in the section "Calculating Head and Pressure Losses Due to Friction Losses."

Step 3. Calculate the pressure change due to the elevation difference along the lateral (see the section "Calculating Pressure Changes Due to Elevation Differences").

Step 4. Calculate the required pressure at the lateral inlet using

$$P_i = P_d + P_f \pm P_e$$

Step 5. If the difference between the design pressure and the inlet pressure exceeds the allowable pressure loss, then change the lateral design using different lengths, diameters, or flow rates.

Step 6. Repeat steps 1 through 4 as necessary until the pressure loss is acceptable.

Example

Calculate the inlet pressure of a lateral for the following design information:

- design pressure = 40 psi
- sprinkler discharge rate = 6 gpm
- number of sprinklers = 33
- lateral length = 1,320 feet
- lateral diameter = 4 inches
- elevation change = 10 feet (uphill flow)

Step 1. Calculate lateral inflow rate.

$$Q = 33 \text{ sprinklers} \times 6 \text{ gpm per sprinkler} = 198 \text{ gpm}$$

Step 2. Calculate the head and pressure loss due to friction for a multiple outlet situation (see section "Calculating Head and Pressure Losses Due to Friction Losses"). Use the Hazen-Williams equation.

$$H_t = [0.36 \times 1050 \times (198 \div 130)^{1.852}] \times (1,320 \div 100) \div 4^{4.87} = 12.7 \text{ feet or } 5.5 \text{ psi}$$

Step 3. Calculate the pressure change due to the elevation difference for uphill flow.

$$P_e = 10 \text{ feet} \div 2.31 = 4.3 \text{ psi}$$

Step 4. Calculate the inlet pressure.

$$P_i = 40 \text{ psi} + 5.5 \text{ psi} + 4.3 \text{ psi} = 49.8 \text{ psi}$$

Step 5. The pressure change along the lateral is 9.8 psi (49.8 psi – 40 psi), which is slightly larger than an allowable pressure loss of 20 percent or 8 psi.

Flow-Control Nozzles

Flow-control nozzles can improve the uniformity of sprinkler discharge rates where excessive pressure losses occur in the sprinkle irrigation system. These nozzles contain a flexible orifice that changes its cross-sectional area as the pressure changes. Under high pressure the area is smaller than under low pressure. Figure 13 shows a diagram of a flow-control nozzle for high and low pressures.

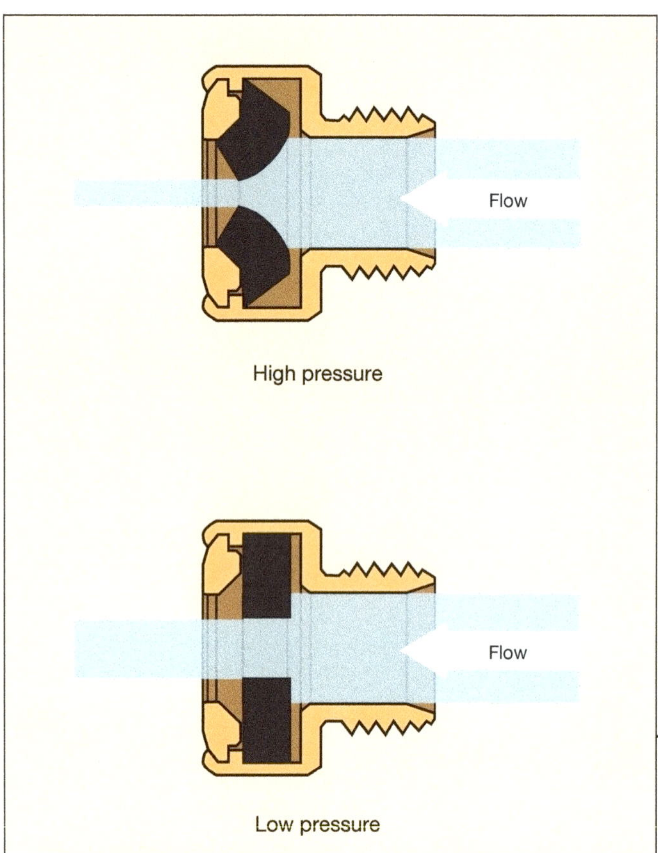

Figure 13. Diagram of a flow-control nozzle under low and high pressure. *Source:* Courtesy of Nelson Irrigation Corporation, Walla Walla, WA.

The sprinkler discharge rate depends on pressure and orifice area. Thus, decreasing the orifice area as pressure increases can decrease the effect of pressure variability on sprinkler discharge rates along a lateral. Figure 14 shows sprinkler discharge rates along the lateral for a flow-control nozzle and a standard nozzle where the pressure varied from 27 psi to 47 psi, primarily due to elevation differences. Smaller changes in discharge rates occurred for the flow-control nozzle as compared with the standard nozzle. The variation of the sprinkler discharge rates of the flow-control nozzles was about 28 percent of that of the standard nozzles.

Flow-control nozzles are rated by their discharge rate instead of the nozzle size. A nozzle with the proper discharge rate must be selected to adequately irrigate the field. One manufacturer offers flow-control nozzle sizes ranging from 4 to 8 gallons per minute.

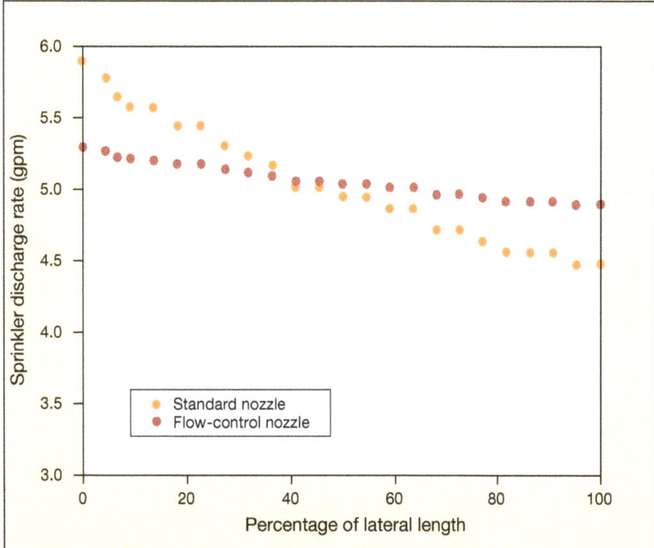

Figure 14. Sprinkler discharge rate in gallons per minute along a lateral for a flow-control nozzle and a standard circular nozzle. The pressure varied from 47 psi at the lateral inlet to 27 psi at the end of the lateral.

Patterns of Applied Water

Patterns of water applied by impact sprinklers depend on sprinkler head characteristics (nozzle sizes, trajectory, rotation rate), pressure, sprinkler spacing, and wind speed and direction. Crop interference and sprinkler riser height are also factors.

Patterns of water application were determined for different operating conditions—using three-dimensional wirescreen plots—by measuring the amount of water caught in catch cans during sprinkler evaluations. Both amount of water and can locations were entered into a computer graphics program, which then developed three-dimensional images of the applied water pattern. The height of the image above the base at any particular location represents the amount of water applied at that location. These three-dimensional images provide a picture of water application patterns for various nozzle sizes, pressures, and wind speeds (figs. 15–21). Each figure shows the applied water pattern of a single sprinkler and of overlapped patterns of several typical sprinkler spacings.

Low pressure (30 psi) caused a ridge of higher water applications near the edge of the pattern around a single sprinkler, called a doughnut-shaped pattern (fig. 15). The average wind speed was 2.5 miles per hour. High water applications also occurred near the sprinkler. Considerable variability occurred in the overlapped pattern for a 30-foot spacing along the lateral and a 40-foot spacing along the mainline or submain, with higher water applications near the sprinklers and small applications in the middle of the pattern. The DU was 72.7 percent.

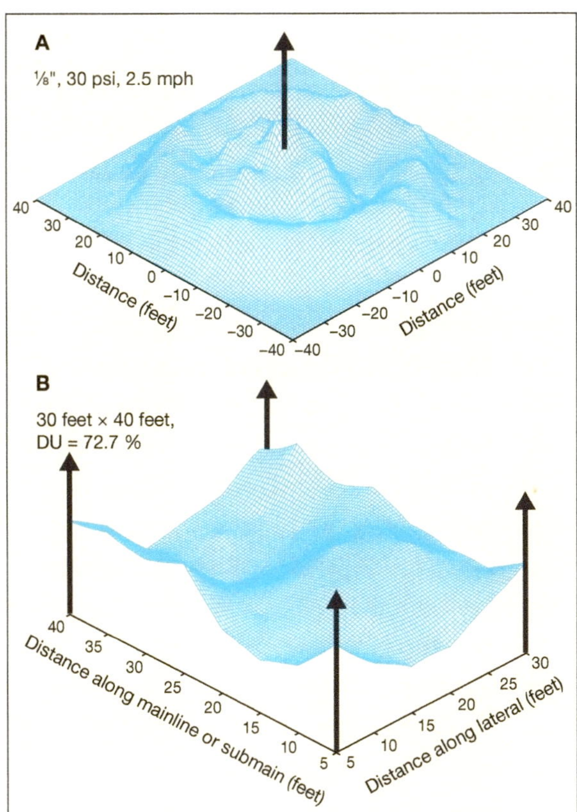

Figure 15. Water application pattern of a ⅛-inch sprinkler operating at 30 psi under a wind speed of 2.5 miles per hour for (A) a single sprinkler and (B) an overlapped pattern for a 30-by-40-foot spacing. The arrows indicate the location of the sprinklers.

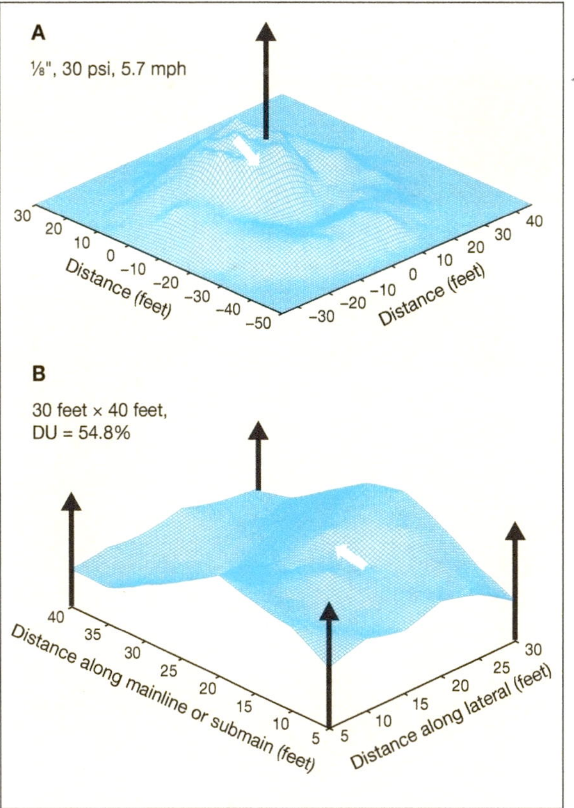

Figure 16. Water application pattern of a ⅛-inch sprinkler operating at 30 psi under a wind speed of 5.7 miles per hour for (A) a single sprinkler and (B) an overlapped pattern for a 30-by-40-foot spacing. The arrows indicate the location of the sprinklers.

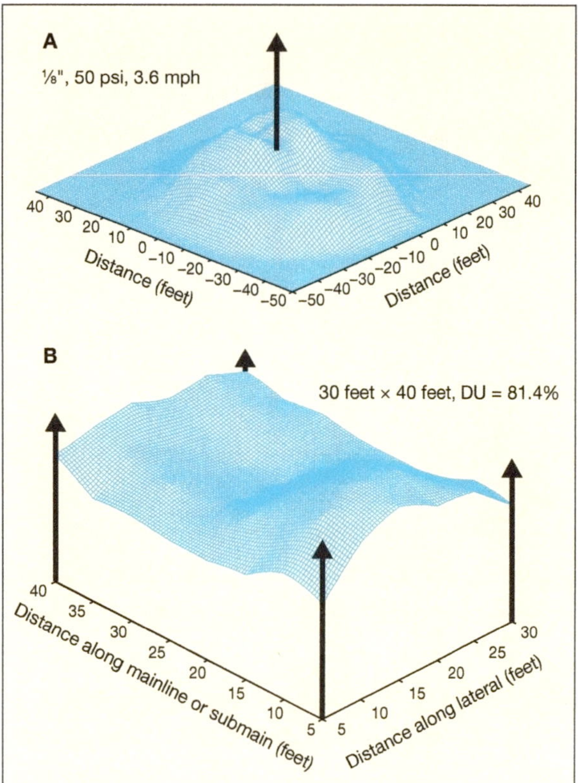

Figure 17. Water application pattern of a ⅛-inch sprinkler operating at 50 psi under a wind speed of 3.6 miles per hour for (A) a single sprinkler and (B) an overlapped pattern for a 30-by-40-foot spacing. The arrows indicate the location of the sprinklers.

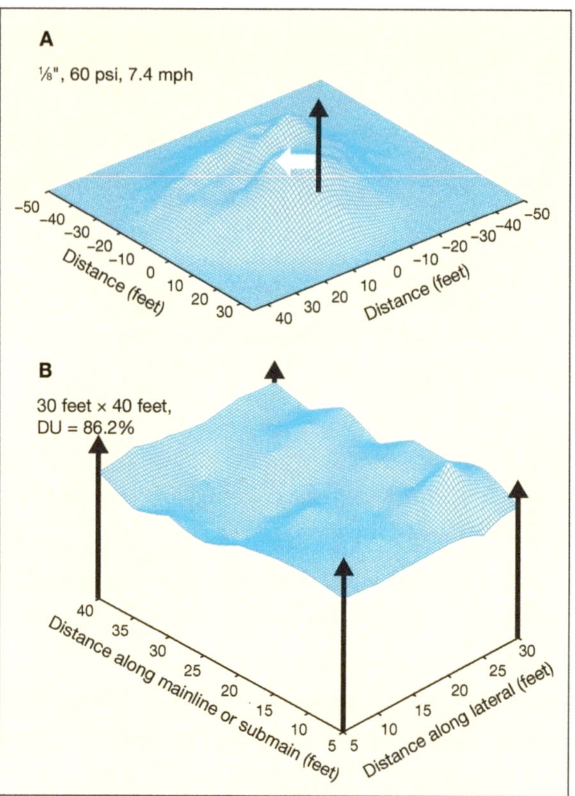

Figure 18. Water application pattern of a ⅛-inch sprinkler operating at 60 psi under a wind speed of 7.4 miles per hour for (A) a single sprinkler and (B) an overlapped pattern for a 30-by-40-foot spacing. The black arrows indicate the location of the sprinklers. The white arrow shows the direction of the wind.

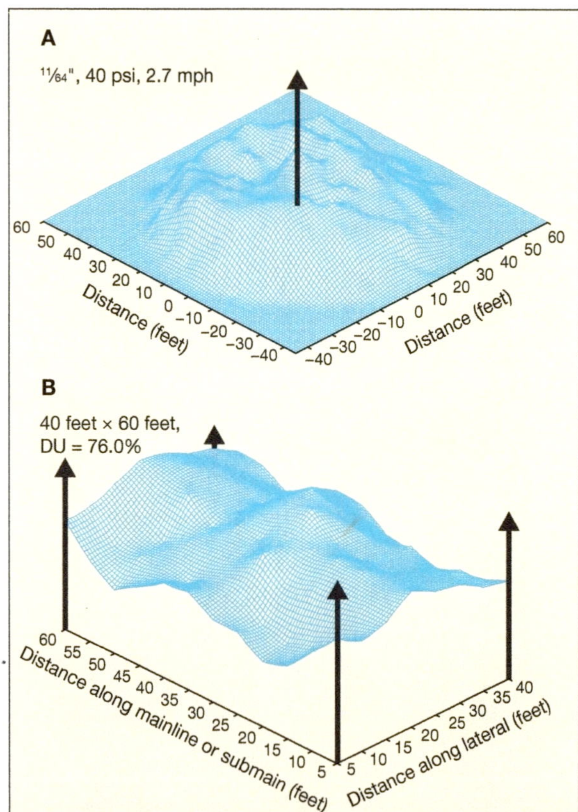

A wind speed of 5.7 miles per hour considerably changed the water application pattern around the single sprinkler (fig. 16). Downwind of the sprinkler, the ridge was still evident, but no ridge occurred upwind. A different overlapped pattern also occurred, with relatively high water applications over about half of the wetted area. The DU was 54.8 percent, considered to be poor.

A pressure of 50 psi resulted in a mound near the sprinkler (wind speed = 3.6 miles per hour) with gradually decreasing water applications with distance from the sprinkler (fig. 17). The higher pressure resulted in a more uniform overlapped pattern (DU = 81.4%)

Figure 19. Water application pattern of an 11⁄64-inch sprinkler operating at 40 psi under a wind speed of 2.7 miles per hour for (A) a single sprinkler and (B) an overlapped pattern for a 40-by-60-foot spacing. The arrows indicate the location of the sprinklers.

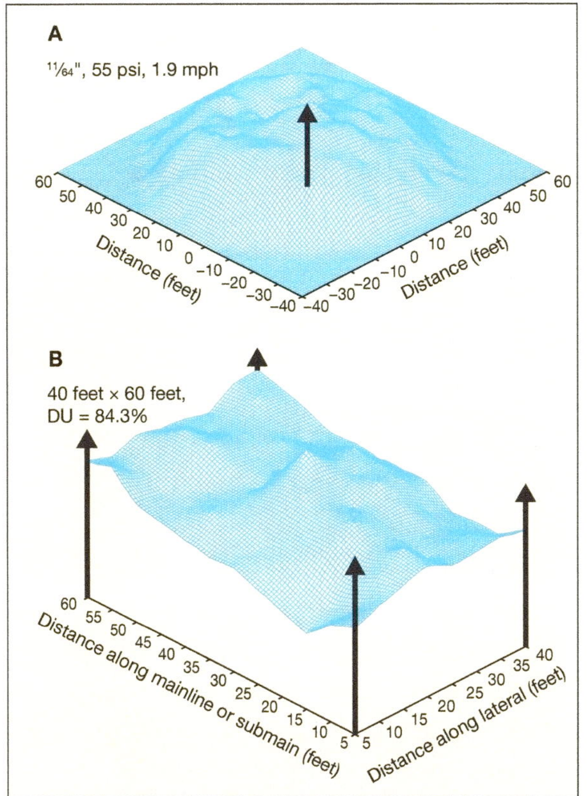

Figure 20. Water application pattern of an 11⁄64-inch sprinkler operating at 55 psi under a wind speed of 1.9 miles per hour for (A) a single sprinkler and (B) an overlapped pattern for a 40-by-60-foot spacing. The arrows indicate the location of the sprinklers.

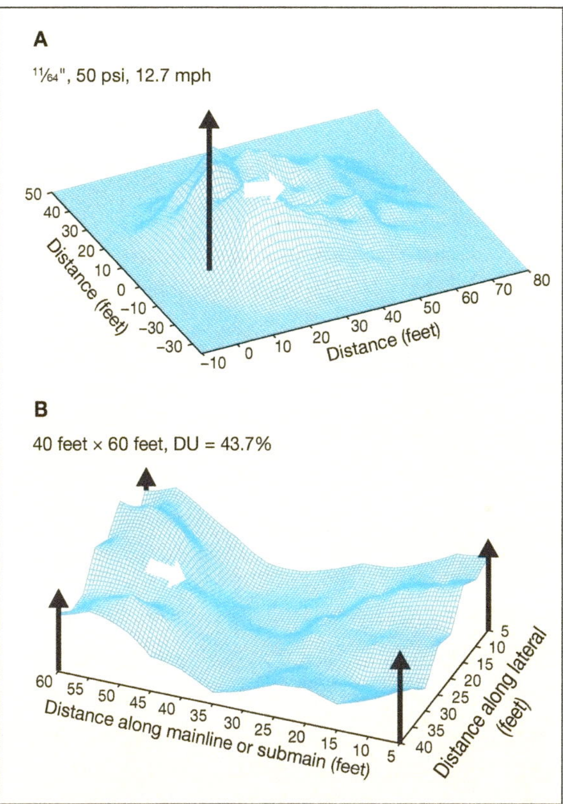

Figure 21. Water application pattern of an 11⁄64-inch sprinkler operating at 50 psi under a wind speed of 12.7 miles per hour for (A) a single sprinkler and (B) an overlapped pattern for a 40-by-60-foot spacing. The black arrows indicate the location of the sprinklers. The white arrow indicates the direction of the wind.

compared to the 30 psi pattern (see fig. 15). Similar patterns occurred for a wind speed of 7.4 miles per hour and a pressure of 60 psi (fig. 18). The DU was 86.2 percent for a 30-by-40-foot spacing.

A slight ridge occurred in the pattern for a 11⁄64-inch single sprinkler operating at 40 psi under low wind conditions (fig. 19). The DU of the overlapped pattern for a 40-by-60-foot spacing was 76.0 percent. Increasing the pressure to 55 psi improved both the single sprinkler pattern and the overlapped pattern, resulting in a DU of 84.3 percent (fig. 20).

Under very high wind conditions, small water applications occurred upwind of the sprinkler while large applications occurred downwind (fig. 21). The high wind created an oval-shaped pattern by reducing the wetted distance perpendicular to the wind direction. The overlapped pattern showed most of the water to be applied in the first half of the wetted pattern, resulting in a poor DU of 43.7 percent.

Below are some key points illustrated by these figures:

- Low pressure can cause poor distribution uniformity of an overlapped water application pattern.
- Increasing the pressure can improve the distribution uniformity.
- Wind greatly distorts the water application pattern of a single nozzle, which in turn results in a poor distribution uniformity of the overlapped pattern.
- Even for high uniformity conditions, considerable variability in water application can occur between sprinklers.

Spacing Guidelines

Overlap of single sprinkler patterns is necessary to uniformly apply water with a sprinkle irrigation system. Criteria for selecting a particular spacing include desired uniformity of application, cost of system, application rate, and irrigation set time. Trade-offs exist between these factors. High uniformity occurs for small spacings, but system and management costs may be high. High application rates, a result of small spacings, require frequent irrigation set changes due to smaller irrigation set times, and high application rates could potentially cause substantial surface runoff. Larger spacings may decrease the uniformity and also system and management costs but potentially at a cost of reduced yield.

The effect of sprinkler spacing on uniformity of the applied water between sprinklers also depends on the water application pattern of a single nozzle (Chen and Wallender 1984). For rectangular sprinkler spacings, the uniformity was greater than 80 percent for sprinkler spacings smaller than about 80 percent of the wetted diameter for single-nozzle water application patterns similar to that shown in figure 8C. Sprinkler spacings smaller than about 70 percent of the wetted diameter were necessary for a uniformity greater than 80 percent for a doughnut-shaped pattern (fig. 8B).

Spacing options include rectangular spacings, square spacings, and equilateral triangular spacings. Spacings for sprinkle irrigation systems used for row and field crops are mostly rectangular, with the larger spacing along mainlines and submains. Square spacings are rarely used for row and field crops. Triangular spacings are not practical for periodic-move systems, but they are commonly used for undertree and landscape sprinkle systems.

General Spacing Recommendations

Spacing recommendations based on the effective wetted diameter of a single sprinkler are shown in table 17. The effective wetted diameter is 90 percent of the wetted diameter listed by manufacturers in their catalogs. Spacings calculated using the recommendations in table 17 will need to be adjusted for compatibility with available pipe section lengths (i.e., 20, 30, or 40 feet).

Example

What spacings are recommended for a ⅛-inch nozzle operating at 50 psi? The manufacturer's wetted diameter is 81 feet. The effective wetting diameter is 0.9 × 81 feet = 73 feet. Using the recommendations in table 17, the respective sprinkler spacing along laterals and lateral spacing along mainlines and submains are as follows: for wind speed smaller than 3 miles per hour, spacings are 0.4 (40%) × 73 feet = 29 feet and 0.65 (65%) × 73 feet = 47 feet; for wind speeds between 4 and 7 miles per hour, spacings are 0.4

Table 17. General recommended sprinkler spacings

Wind speed (mph)	Sprinkler spacing along lateral (% of the effective wetted diameter)	Lateral spacing along mainline/submain (% of the effective wetted diameter)
0–3	40	65
3–7	40	60
7–10	40	60
> 10	30	50

Sources: Rain Bird Corporation 1971; Keller and Bliesner 1990.
Note: These recommendations are expressed as a percentage of the effective wetted diameter and wind speed. The effective wetted diameter is 90% of the manufacturer's wetted diameter. The effective diameter is used because the last few feet of the wetted diameter contribute little to the soil moisture replenishment during irrigation.

(40%) × 73 feet = 29 feet and 0.6 (60%) × 73 feet = 44 feet; for wind speeds greater than 10 miles per hour, spacings are 0.3 (30%) × 73 feet = 22 feet and 0.5 (50%) × 73 feet = 36 feet. A sprinkler spacing of 30 feet and a lateral spacing of 40 feet should be used to account for both moderate wind speeds and the available pipe section sizes.

Spacing—Uniformity Relationships

The bar graphs of figures 22 through 25 illustrate the effect of sprinkler spacing on distribution uniformity for several nozzle sizes and wind speeds. These relationships, developed from field evaluations, are appropriate for hand-move, wheel-line, and portable solid-set sprinkle irrigation systems. Spacings along the mainline and submain are shown along the bottom of the figures. For each mainline spacing, a series of colored bar graphs shows the distribution uniformity for different sprinkler spacings along the lateral. Each color represents a particular sprinkler spacing. The left-hand side of the figure shows the distribution uniformity (DU).

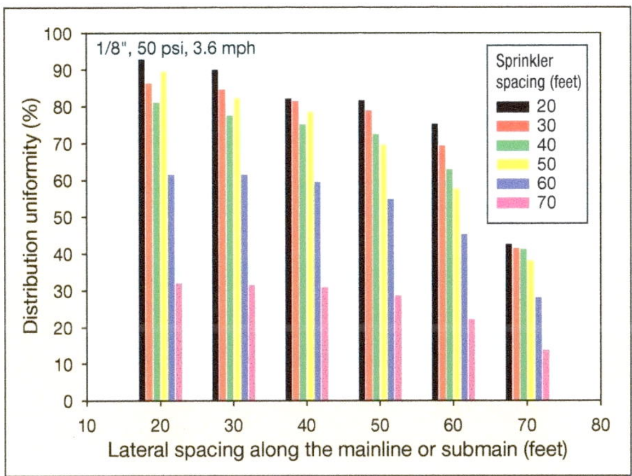

Figure 22. Relationship between distribution uniformity and sprinkler spacings for a ⅛-inch nozzle operating at 50 psi under a wind speed of 3.6 miles per hour.

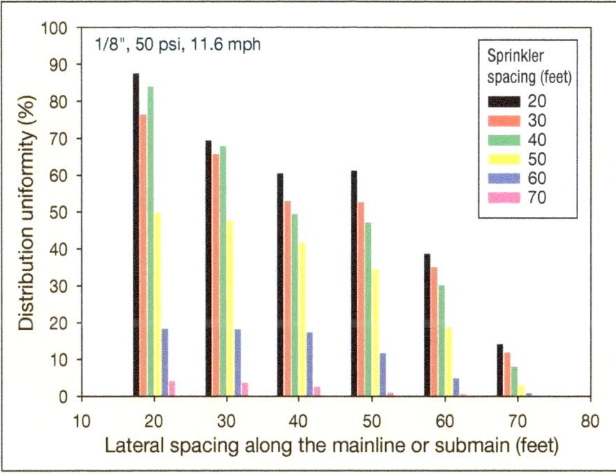

Figure 23. Relationship between distribution uniformity and sprinkler spacings for a ⅛-inch nozzle operating at 50 psi under a wind speed of 11.6 miles per hour.

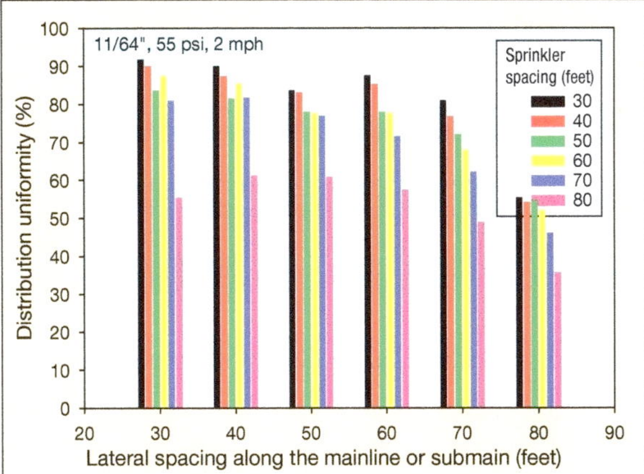

Figure 24. Relationship between distribution uniformity and sprinkler spacings for an 11⁄64-inch nozzle operating at 55 psi under a wind speed of 2 miles per hour.

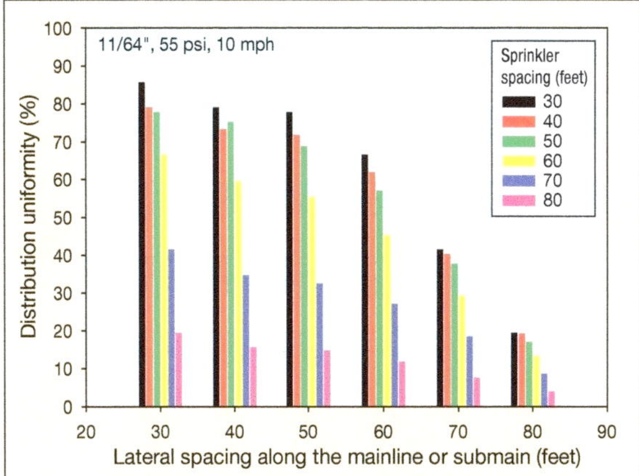

Figure 25. Relationship between distribution uniformity and sprinkler spacings for an 11⁄64-inch nozzle operating at 55 psi under a wind speed of 10 miles per hour.

These bar graphs generally show that under low wind conditions, good uniformity is provided by spacings of 30 feet by 40 feet for 1/8-inch nozzle sizes and spacings of 40 feet by 60 feet for 11/64-inch nozzles. These spacings are commonly used for these nozzle sizes. Smaller spacings may slightly increase the DU but at higher system and labor costs. In this study, however, high wind conditions greatly decreased the DU for these spacings. Under these wind conditions, small spacings improved the DU, but these spacings may not be practical because of increased costs.

Figure 26. Effect of high wind conditions on the distribution of water. Note that little irrigation water is applied upwind of the sprinkler lateral. The dark soil is that wetted by the irrigation. *Photo:* Larry Schwankl.

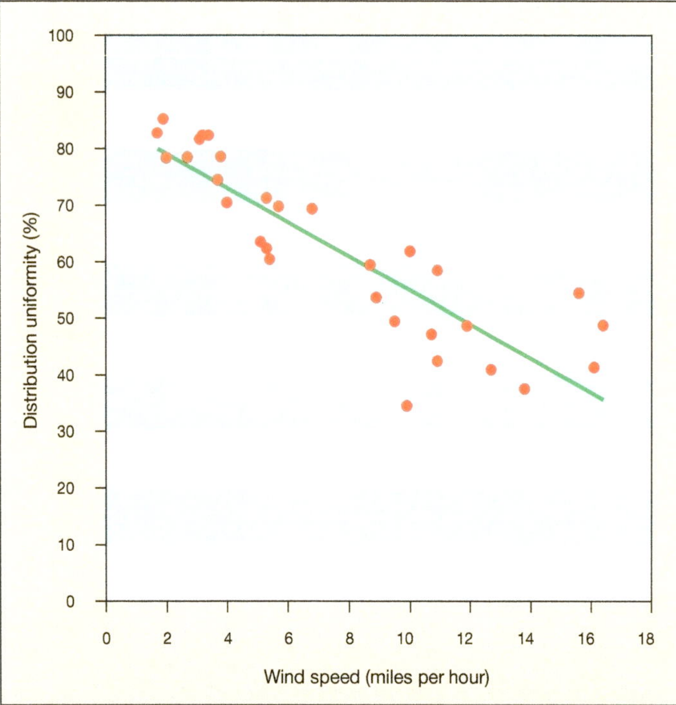

Figure 27. Relationship between the distribution uniformity (DU) and wind speed in miles per hour, developed from field catch-can data. The solid line shows that the distribution uniformity decreased by an average of 3 percentage points for each 1 mile per hour increase in wind speed.

Coping with Wind Effects on Uniformity

Wind is the major factor contributing to poor uniformity of the water applied by sprinkle irrigation systems. As discussed earlier, wind distorts the pattern of applied water by blowing much of the spray downwind of the sprinkler (fig. 26). Wind speed decreases the distribution uniformity (DU) in a direct manner, based on numerous catch-can evaluations (fig. 27). Good uniformity occurs until wind speeds exceed about 5 miles per hour. For every 1 mile per hour increase in wind speed, the DU decreases by an average of 3 percentage points.

Orienting sprinkler laterals perpendicular to the wind direction to improve the DU under windy conditions is commonly recommended. However, field experiments by the authors and by Kasapligil (1990) have shown similar DUs to occur for different lateral orientations (perpendicular or parallel) relative to the wind direction, but the water application patterns differed greatly. A wind direction perpendicular to the lateral resulted in little water applied upwind of the sprinkler lateral, with most of the water applied within about 40 feet downwind; however, a wind direction parallel to the sprinkler line resulted in high water applications along the sprinkler lateral and little water applied in the middle of the pattern (fig. 28).

In general, few measures exist that are practical for improving distribution uniformity under relatively high wind speeds. While decreasing sprinkler spacings can improve the DU, the reduced spacings may not be very practical both in terms of management and cost. The adjustment in sprinkler spacings needed to offset the effect of high wind speed on distribution uniformity is illustrated using figure 25. The DU was 62 percent for a wind speed of 10 miles per hour for a 40-by-60-foot spacing. Spacings of 30 by 30 feet are needed to increase the DU to over 80 percent.

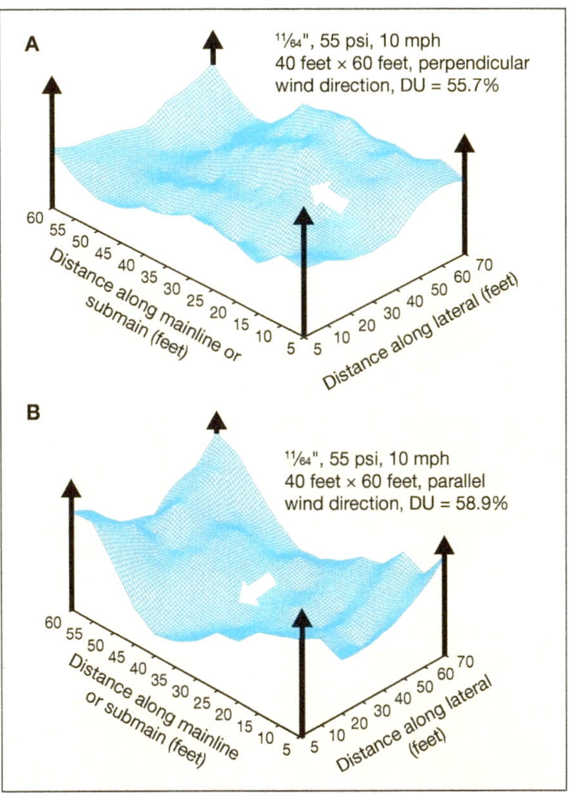

Figure 28. Water application pattern of an 11⁄64-inch sprinkler operating at 55 psi under a wind speed of 10 miles per hour for (A) a wind direction perpendicular to the sprinkler lateral and (B) a wind direction parallel to the sprinkler lateral. The sprinkler spacing is 40 by 60 feet. The black arrows indicate the location of the sprinklers. The white arrows indicate the direction of the wind.

Measures to Reduce Wind Effects on the DU

- Offset the sprinkler laterals. Offsetting irrigation sets can improve the field-wide DU by 10 to 20 percentage points compared with the DU of normal sets. This measure involves placing the sprinkler laterals of the current irrigation set midway between the lateral locations of the previous irrigation set. For the next irrigation set, the laterals are placed at the same locations as the first irrigation set. This offsetting continues throughout the irrigation season.
- Select a low-trajectory sprinkler head. Wind speed increases with height above the ground surface. A low-trajectory sprinkler head reduces the height of the spray pattern, but it also decreases the wetted diameter, which could reduce uniformity. Little information exists on the effect of smaller trajectories on uniformity.
- Reduce the pressure. Low pressure results in relatively large sprinkler droplets, which are less susceptible to wind compared with smaller droplets.
- For wind conditions that are consistently high, it may be necessary to convert to an irrigation system that is not affected by wind, such as drip irrigation.

Evaluating and Improving Sprinkle System Performance

Sprinkle systems should be evaluated periodically to monitor the irrigation system performance, ensuring that adequate water is being applied and that the system performance is satisfactory. The system performance can change due to pump wear, declining groundwater levels, clogged screens, worn sprinkler nozzles, changes in nozzle sizes, changes in the sprinkle irrigation system, and so on.

Figure 29. Measuring the sprinkler nozzle discharge rate using a hose and a 5-gallon bucket. *Photo:* Blaine Hanson.

Evaluating Sprinkle Irrigation System Performance

Measuring the Sprinkler Discharge Rate and Application Rate

Measure the sprinkler discharge rates to determine the application rate. Equipment needed will be a 3- to 4-foot section of garden hose, a container, a stopwatch, and a method of measuring the volume of water in the container.

Step 1. Measure the nozzle flow rate using a section of garden hose and a container. Insert one end of the hose over the nozzle and direct the other end into the container (fig. 29). Measure the time it takes for water to fill the container. Convert the time into minutes.

Step 2. Determine the number of gallons of water in the container.

Step 3. Calculate the nozzle discharge rate. Divide the gallons caught by the time expressed in minutes.

Step 4. Calculate the application rate (AR) in inches per hour using the measured sprinkler discharge rate in gallons per minute (q), sprinkler spacing along lateral (S_l) and mainline (S_m) in feet, and the following equation:

$$AR = (96.3 \times q) \div (S_l \times S_m)$$

Calculating the Amount of Applied Water

Calculate the depth of applied water using

$$D = AR \times T$$

where:

D = the depth of applied water in inches

T = the irrigation set time in hours

AR = the application rate in inches per hour

Compare this depth with the desired depth of application.

Checking the Pressure

Periodically check the pressure. Monitor the pressure at the pump with a pressure gauge.

The pressure can also be checked by measuring the nozzle pressure of the sprinklers using a pitot gauge (fig. 30). This device is a short section (about 1 inch long) of brass tubing attached to a fitting. The brass tubing is bent at a 90-degree angle. A standard pressure gauge is attached to the fitting. The end of the brass tubing is inserted into the nozzle orifice, and the pressure is read from the pressure gauge.

Figure 30. Pitot gauge used for measuring sprinkler nozzle pressure. *Photo:* Blaine Hanson.

Note that inserting the pitot gauge into the nozzle orifice will plug the nozzle and cause the discharge rate of the other nozzles to increase slightly. However, this increase is very small for systems with a large number of sprinklers, typical for most sprinkle irrigation systems, and thus the effect of plugging one nozzle is negligible on irrigation system flow rate. For systems with a few sprinklers (less than 10), the pitot gauge should be held about ⅛ inch from the nozzle. The problem with this procedure is that it is difficult to read the pressure gauge because of the spray from the water jet, and slight variations in the position of the tip of the pitot gauge in the jet can affect the reading.

Other Items to Consider

Other items to check include the following:

- Leaks. Water loss from leaks can be considerable (fig. 31). The water loss from leaky drain valves equals about 30 percent of the sprinkler discharge rate for one wheel-line system.
- Malfunctioning sprinkler heads. This includes leaks and improper rotation.
- Worn nozzles. Sprinkler nozzles become worn over time, especially when the irrigation water contains sand or other particles. Nozzle wear increases the application rate, potentially causing over-irrigation, behavior which may not be apparent to the irrigator. Nozzle wear can be checked using drill bits of the same diameter as the initial nozzle diameter. Nozzle wear may not affect the application uniformity, but the uniformity could be affected by sprinkler head wear. A study of age effects on catch-can uniformity showed a DU of 81.6 percent for new nozzles, 79.1 percent for 2-year-old nozzles, and 74.8 percent for nozzles older than 3 years.

Figure 31. Flow rates of leaks in a wheel-line sprinkler lateral: 7.3 gallons per minute (A) and 11.1 gallons per minute (B). *Photos:* Steve Orloff.

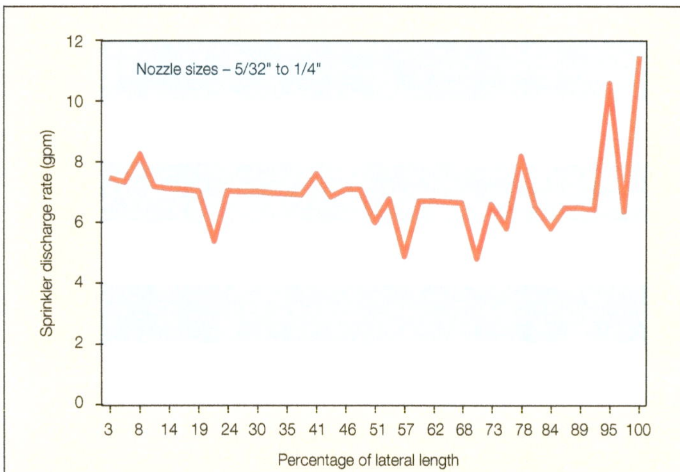

Figure 32. Effect of mixed nozzle diameters along the lateral length. Sprinkler discharge rates ranged from about 5 gallons per minute to 11.5 gallons per minute due to mixed nozzle diameters.

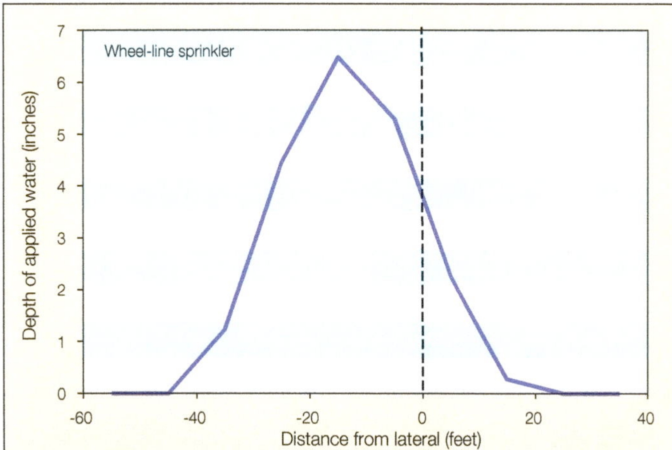

Figure 33. Effect of nonvertical sprinkler riser on the distribution of applied water. The wetted distance was about 25 feet on one side of the lateral and over 40 feet on the other side due to the nonvertical riser. The dashed line shows the location of the sprinkler lateral.

- Mixed nozzle sizes. Mixing nozzle sizes along sprinkler laterals results in different amounts of applied water along the lateral (fig. 32) and can affect the distribution uniformity.
- Malfunctioning valves and hydrants.
- Clogged screens.
- Clogged nozzles.
- Crop interference with spray pattern. This results from inadequate riser height.
- Risers not vertical or leaning. Riser angles smaller than 10 degrees from the vertical have little effect on the distribution uniformity of the applied water under low wind conditions. A riser angle of 20 degrees can reduce the distribution uniformity by 7 to 10 percentage points under low wind conditions. Figure 33 shows a greater wetted distance on one side of a sprinkler lateral than on the other side due to a nonvertical riser.
- Excessive surface runoff. This indicates that the application rate is too high.
- Different day and night set times.

Evaluating Catch-Can Uniformity

Changes in the existing irrigation system that deviate from the initial design conditions can change the uniformity of applied water. A catch-can test can help evaluate the effect of the changes on the pattern of applied water. The catch-can test must be done under low wind conditions. High wind conditions will mask the effect of changes in the irrigation system on the distribution uniformity. The procedure for conducting catch-can tests is in the section "Measuring the Catch-Can Distribution Uniformity."

Improving Sprinkle Irrigation System Performance

Economic incentives for improving the performance of the sprinkle irrigation system are increased yields and reduced energy costs due to decreased water applications. More uniform water applications can potentially increase yields and thus revenue. However, predicting the effect of more uniformly applied water on crop yield is difficult because of factors discussed in the section "Uniformity of Periodic-Move Sprinkle Systems." Water applications and energy costs can be reduced by decreasing the irrigation set time. However, reduced energy costs only may be insufficient to offset the costs of an improvement such as changing sprinkler spacings. One study showed that the effect of wind speed on the uniformity of applied water has a significant effect on the economic performance of a sprinkle irrigation system (Brennan 2008). The study also showed that the larger the difference between the existing irrigation system and an improved system, the larger the economic incentive is. Distribution uniformities smaller than 70 percent under low wind conditions indicate a potential for improving the irrigation system performance. Measures for improving system performance include the following:

Coping with Wind Effects

- Irrigate during low wind conditions such as at night or early morning. This requires sufficient pump capacity to irrigate only during low wind periods.
- Install sprinkler heads with a smaller trajectory angle. Catch-can tests should be conducted before changing all of the sprinkler heads to determine the effect of this measure on the uniformity.
- Increase the droplet size by increasing the nozzle diameter. However, this approach will increase the flow rate in the pipeline, thus increasing pressure losses due to friction. Catch-can tests should be conducted before changing all of the nozzles.
- Offset the sprinkler laterals by placing the sprinkler laterals of the current irrigation set midway between the lateral locations of the previous irrigation set. For the next irrigation set, the laterals are placed at the same locations as the first irrigation set. This offsetting continues throughout the irrigation season. The offset irrigation sets can improve field-wide DU by 10 to 20 percentage points compared with the DU of normal sets.
- Reduce sprinkler spacings if the existing spacings are excessive compared with normal design practices for a given type of sprinkle irrigation system.

Increasing Pressure

- Adjust pump impeller (semiopen impellers).
- Replace or repair worn pump.
- Replace worn nozzles.
- Install a booster pump.
- Reduce the number of laterals, particularly if the low pressure was caused by expanding the irrigation system without changing the pump. This measure also will increase the pressure if the low pressure is caused by declining groundwater levels.

Increasing Flow Rate

- Repair a worn pump.
- Install a new pump that provides the desired higher pressure and flow rate.
- Change to larger nozzle sizes. Note that this change will also decrease the pressure.

Adjusting for Inadequate Sprinkler Spacings

- Reduce the sprinkler spacing. This measure is generally not practical because of the cost of changing an existing system.
- Select a different sprinkler head and nozzle combination that increases the wetted diameter. Catch-can tests should be conducted to ensure that an improvement in uniformity will occur.
- Increase both pressure and nozzle diameter to increase the wetted area.

Reducing Crop Interference

- Increase riser height for row or field crops.

Reducing Excessive Pressure Variation

- Install flow-control nozzles.

Maintaining the Sprinkle Irrigation System

- Repair or replace malfunctioning sprinkler heads.
- Unclog plugged nozzles.
- Replace worn sprinklers.
- Repair leaks.
- Use same-sized nozzles along the lateral.
- Maintain vertical risers.

Measuring the Catch-Can Distribution Uniformity

The catch-can distribution uniformity is determined by installing a grid of catch cans throughout the area wetted by the overlapped sprinklers, operating the system, and then measuring the volume of water caught in the catch cans. Computer software can be used to calculate the distribution uniformity using the catch-can data.

Why Measure the Catch-Can Distribution Uniformity?

The catch-can distribution uniformity (DU) is an index of how evenly the sprinkler-applied water is distributed throughout the wetted area. A catch-can test is recommended if changes to the sprinkle irrigation system are made that can affect the distribution of water. Such changes include changes in the spacing, different sprinkler heads and nozzles, declining groundwater levels, and pressure changes due to declining pump performance.The catch-can test must be conducted under low wind conditions to evaluate the effect of irrigation system changes. Conducting the test under high wind conditions is of little or no value because the wind is the major contributor to poor uniformity and will mask the effect of system changes on the DU.

Procedure

Hand-Move and Wheel-Line Sprinklers

Step 1. Spread catch cans evenly along a sprinkler lateral line between the location of the current set and the location of the next set (fig. 34). A recommended catch-can spacing is 10 feet. Catch-can diameter should be 4 to 6 inches. Smaller catch-can diameters should be avoided because research has shown that they do not accurately reflect the amount of applied water reaching the ground surface. The same diameter must be used for all catch cans.

Step 2. Measure the amount of water in the catch cans after operating the current set for at least 1 to 2 hours. However, measuring the water caught for the entire set time will give the best results. Leave the cans at the same locations for the next set.

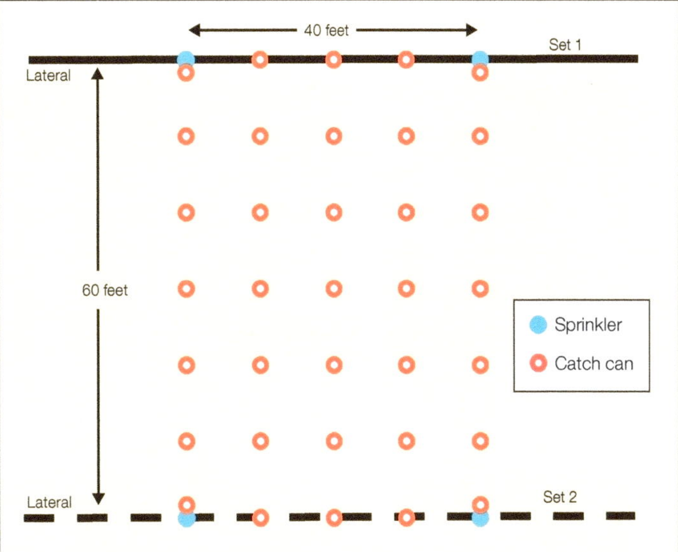

Figure 34. Layout of catch cans for a 40-by-60-foot sprinkler spacing for a hand-move or wheel-line sprinkle irrigation system. Set 1 is the first set of the test; after the end of Set 1, the sprinkler lateral is then moved to Set 2 and the catch-can test repeated.

Step 3. Measure the amount of water in the catch cans after operating the next set. The same time periods must be used for each set of measurements.

Step 4. Add the amount of water caught during each set to determine the total amount of water applied between the sprinklers.

Step 5. Calculate the average amount of water caught for all of the catch cans. The average amount equals the sum of individual amounts caught, divided by the number of catch cans.

Step 6. Calculate the average amount caught in the lowest one-fourth of the catch cans. For example, if 20 catch cans are used, calculate the average amount caught in the 5 cans with the smallest amounts of water. This value is the average of the low quarter (see the section "Uniformity and Efficiency of Sprinkle Irrigation Systems: General Considerations").

Step 7. Calculate the DU by dividing the average of the lowest one-fourth by the average of all of the cans. Multiply by 100 to express the DU as a percentage.

Portable Solid-Set Sprinklers

Step 1. Spread 20 to 30 catch cans evenly between four adjacent sprinklers (fig. 35). Catch-can diameter should be 4 to 6 inches. The same diameter must be used for all catch cans. Because all of the sprinklers will be operated at the same time, make the catch-can measurements of a particular evaluation for one irrigation set only. Install the catch cans where the pressure is the smallest.

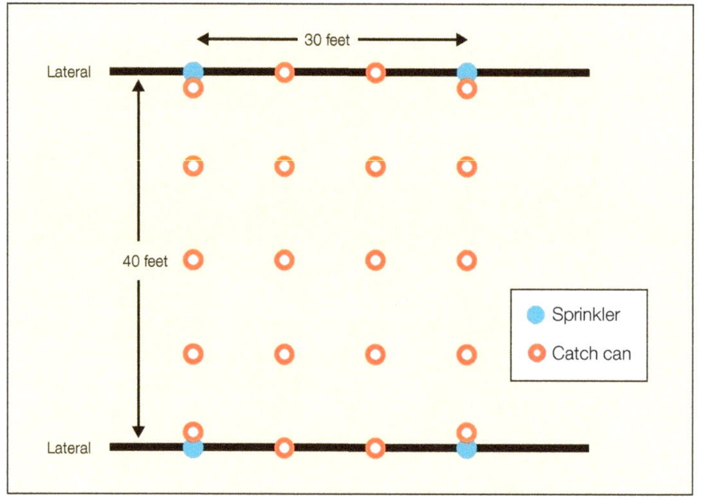

Figure 35. Layout of catch cans for a 30-by-40-foot spacing of a portable solid-set sprinkle irrigation system.

Step 2. Measure the amount of water collected in each catch can after operating the system for at least 1 to 2 hours. For best results, however, measure the amount caught for the set time.

Step 3. Calculate the average amount caught for all catch cans. The average equals the sum of all amounts, divided by the number of catch cans.

Step 4. Calculate the average amount caught in the lowest one-fourth of the catch cans. For example, if 20 cans are used, calculate the average amount in the lowest five catch cans (20 catch cans divided by 4). This value is the average of the low quarter.

Step 5. Calculate the DU by dividing the average of the lowest one-fourth by the average amount. Multiply by 100 to express the DU as a percentage.

Software for Calculating the DU

Using a single sprinkler may be desirable for a catch-can test. The data of the single sprinkler test can be used to determine the distribution uniformity for various sprinkler spacings. The major problem with this approach is overlapping the single sprinkler water pattern to obtain the applied water patterns of the overlapped spacings. Determining the overlapped patterns can be confusing and time-consuming.

The software CATCH-3D, developed by the Biological and Irrigation Engineering Department at Utah State University, can be used to calculate the overlapped patterns and their distribution uniformities. It can also be used for the overlapped tests previously described. The software can be obtained from the Biological and Irrigation Engineering Department at Utah State University at their Web site, http://www.neng.usu.edu/bie/faculty/merkley/BIE6110.htm. It provides a 3-D picture of the pattern of applied water and also calculates the statistics for a single sprinkler and for any desired sprinkler spacing. A manual is also provided.

Continuous-Move Sprinkle Irrigation Systems

Center-Pivot and Linear-Move Sprinkle Irrigation Systems

Center-Pivot Sprinkle Systems

Center-pivot irrigation systems consist of a lateral pipeline mounted on top of self-propelled towers (fig. 36). The lateral is suspended about 10 to 15 feet above the ground. Distance between towers, that is, span length, can range from 90 to 250 feet. A flexible joint connects the spans. A typical lateral length is about 1,300 feet, which can irrigate about 130 acres, for a complete circle. The common lateral diameter is 6⅝ inches, but diameters up to 10 inches also are available. The lateral rotates around a fixed pivot point, with the rate of rotation controlled by the outermost tower.

Because center-pivot systems rotate around a fixed point, the area that is irrigated per unit length of lateral increases as the distance from the pivot point increases. Thus, instantaneous application rates must increase with distance from the pivot point to apply the same depth of water throughout the field and maintain high field-wide uniformity. Application rates are increased by using progressively larger nozzle sizes with distance from the pivot point, progressively smaller sprinkler spacings, or a combination of both. Application rates may exceed several inches per hour near the end of the lateral, whereas application rates are a fraction of this near the center of the pivot.

The management of center-pivots involves selecting the appropriate travel speed so that the desired depth of water can be applied throughout the field during one revolution of the machine. An advantage of this irrigation method is that the traveling irrigation lateral ends at the initial starting

Figure 36. Center-pivot sprinkle irrigation systems in (A) Siskiyou County irrigating alfalfa and (B) Kern County irrigating a row crop. *Photos:* Steve Orloff (A); Blake Sanden (B).

point without moving sprinkler lines, provided that the entire circular field is planted to the same crop and the irrigation is a complete revolution of the pivot.

Center-pivot systems are best suited for soils with higher infiltration rates. In the Central Valley of California, however, soils with low infiltration rates coupled with the field slopes and the relatively high center-pivot application rates can result in considerable surface runoff, which has discouraged center-pivot irrigation in the valley. However, alfalfa irrigators and irrigators in areas with level fields are now using center-pivot irrigation in the Central Valley.

The design of center-pivot systems is complex, requiring the selection of appropriate sprinkler nozzle sizes and spacings along the lateral for high uniformity of applied water. Because of this complexity, computer programs are normally used for their design. More information on design and management can be found in the manuals *Center Pivot Design* (The Irrigation Association 2000) and *Sprinkler Irrigation Systems* (Iowa State University Midwest Plan Services 1999).

Linear-Move Machines

Linear-move sprinkle machines use the same technology as center-pivots, but they travel in a straight line. A ditch or pipeline supplies water to the irrigation system (fig. 37). A guidance system keeps the machine traveling in a straight line. An engine-driven pump mounted on the tower adjacent to the ditch or pipeline supplies water and electrical power to the lateral. These systems are best suited for rectangular fields with no obstructions and a relatively uniform topography. An uneven topography can cause problems with the guidance system. In contrast to center-pivot systems, the instantaneous application rate along the lateral length is relatively constant with the distance along the lateral. This is because sprinkler spacing and discharge rate are constant along the lateral. This system can be used on soils with low infiltration rates.

Figure 37. Ditch supplying water to a linear-move sprinkle irrigation system. *Photo:* Blaine Hanson.

Irrigation water is supplied to the linear-move machine by either a ditch or a pipeline. For the pipeline supply, a flexible hose is used to supply water, with one end attached to a riser on the water supply line and the other end attached to the linear-move machine. The irrigation system can be operated continuously using the ditch system, but it must be periodically stopped to move the flexible hose from one riser to another for the pipeline supply systems.

Several options are suggested for the management of linear-move systems. The main management problem is that, unlike center-pivot systems, the irrigation system does not end up at the initial starting point of the irrigation at the end of an irrigation set. Thus, the machine must be moved back to the starting point for the next irrigation to avoid over-irrigating the most recently irrigated part of the field. One option is to irrigate the entire field and then move the machine in the "dry" mode back to the starting point of the irrigation. This may require a "rest" time to dry the soil of the most recently irrigated part of the field sufficiently to move the towers over the soil without getting them stuck. Another option is to irrigate the first half of the field, run the machine over the last half in the "dry" mode to the end of the field, then reverse the direction and irrigate the last half of the field, then run the machine in the "dry" mode over the first half of the field to the initial starting point. For this option, the "dry" mode should consist of increasing the speed of the machine to apply a small application of water that will not wet the soil enough to cause trafficability problems while moving back across the field. This approach will allow the water to continue to flow in the supply ditch or pipeline during the "dry" mode.

Sprinklers

Many sprinkler options exist for center-pivot and linear-move systems. These include impact sprinklers operating from low pressures (30 psi) to high pressures (more than 60 psi) and many types of spray nozzles with different wetted diameters, nozzle sizes, and deflector plates that are stationary, rotating, or oscillating. Spray nozzles operate at low pressures of 10 to 30 psi and are frequently mounted on drop tubes, which lower the sprinkler heads below the lateral pipe. The drop tube approach is now commonly used.

In addition, these systems may also be classified as low energy precision application (LEPA), low elevation spray application (LESA), and mid-elevation spray application (MESA). LEPA systems apply water directly to the soil surface using drop tubes without sprinkler nozzles that discharge the water into the furrow. Furrow dikes have been used to prevent surface runoff. Spray nozzles of LESA systems are about 1 to 2 feet above the ground. MESA nozzles are about 5 to 10 feet above the ground.

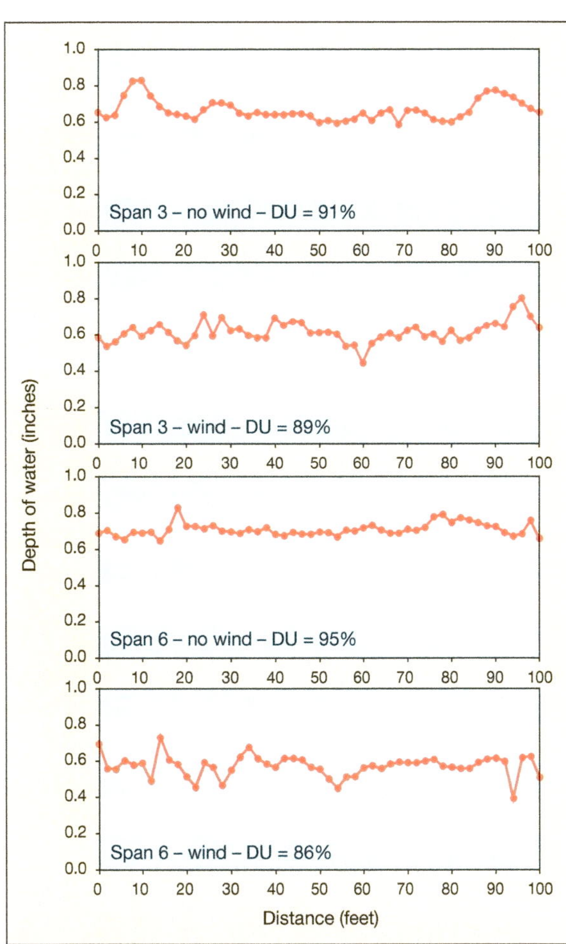

Figure 38. Water application patterns across span 3 (middle span) and span 6 (end span) of a center-pivot sprinkle irrigation system for low and high wind conditions.

Uniformity

Higher distribution uniformities with potential uniformities of 80 to more than 90 percent can occur for center-pivot and linear-move sprinkle systems. The fairly continuous movement of these machines maintains a better water application pattern even under windy conditions. This is illustrated in figure 38, which shows catch-can data across the middle and end spans of a center-pivot system for low wind speeds (morning irrigation) and high wind speeds (afternoon irrigation). Spray nozzles with a rotating deflector pad were used for both spans. Little difference in amounts of applied water and DU were found when comparing the wind conditions.

Traveling Sprinkle Irrigation Systems

Traveling sprinkle irrigation systems, sometimes referred to as traveling guns, big guns, or hose-reel systems, are not commonly used in California except for pasture irrigation of some coastal and mountain areas and also for disposal of dairy lagoon water. An advantage of these systems is that, when irrigating a field divided into smaller sections used for grazing dairy cows, all sections of the field can be irrigated without having to irrigate each section separately. This would result in a potentially higher uniformity and smaller labor cost compared with surface irrigation methods.

Description

Traveling sprinkle irrigation systems consist of a sprinkler carriage, hose carriage, and a flexible hose. These systems can move continuously down the field (fig. 39) or can be moved manually (fig. 40), depending on the system configuration. A strip or lane is irrigated as the sprinkler carriage moves across the field length.

Two methods generally are used for the continuous-move system. The cable tow system uses a cable to pull the sprinkler carriage across the field. A flexible hose supplies irrigation water to the sprinkler. The cable is anchored at one end of the field and a power winch pulls the cable, flexible hose, and sprinkler carriage across the field. The hose is dragged along the ground surface as the sprinkler carriage moves, and then it is reeled up after the sprinkler reaches the end of the strip.

Another approach uses a hard, semiflexible hose to move the sprinkler carriage. One end of the hose is attached to a hose reel carriage, and the other end is connected to the sprinkler carriage. The hose reel carriage slowly reels in the hose, which in turn pulls the sprinkler carriage down the field. Prior to an irrigation set, the hose and sprinkler carriage are pulled across the field with a tractor. At the end of the irrigation set, the sprinkler and hose carriages are manually moved to a new set.

These continuous-move systems generally use a sprinkler carriage containing one large sprinkler head with a nozzle size of 1 inch or larger and an operating pressure ranging from 80 to 100 psi. The sprinkler head can rotate 360 degrees or some fraction thereof, usually 230 to 270 degrees. These systems can irrigate a strip or lane about 300 to 400 feet wide during an irrigation set.

Figure 39. Sprinkler carriage of a continuous move, hose-reel, traveling sprinkler, or big gun irrigation system. *Photo:* Larry Schwankl.

Figure 40. Sprinkler carriage of a small, manually moved, traveling sprinkler. *Photo:* Steve Orloff.

Depth of Applied Water

The average depth of water applied by traveling sprinkle systems depends on the sprinkler discharge rate, spacing between lanes, and the travel speed. It can be calculated using the following equation:

$$D = (1.605 \times Q) \div (W \times S)$$

where:

D = the average depth of applied water in inches

Q = the discharge rate of the sprinkler in gallons per minute

W = the lane spacing in feet

S = the travel speed in feet per second

1.605 = the constant used only for these units of water, discharge rate, lane spacing, and speed

Uniformity

The uniformity of the applied water depends on the water application pattern of the single sprinkler, which in turn depends on nozzle discharge rate, rotation rate of sprinkler, angle of rotation, nozzle trajectory, and wetted diameter. It also depends on lane spacing, travel speed, and wind speed and direction. Evaluations of three traveling guns showed different nonoverlapped water application patterns (fig. 41). System MC3, which rotated about 270 degrees, had a very nonuniform pattern across the land with peaks at about 75 to 80 feet from the sprinkler and a valley at the sprinkler location. Reasons for this behavior were not clear. System BR1 had a relatively uniform pattern until near the edge of the wetted area, while System HR3 had a nonuniform pattern caused by a wind direction perpendicular to the lane, which resulted in more water applied downwind of the sprinkler. A different relationship between DU and lane spacing occurred among the three sprinkle systems (fig. 42). For new installations, the manufacturer should be consulted concerning the appropriate lane spacings for their systems.

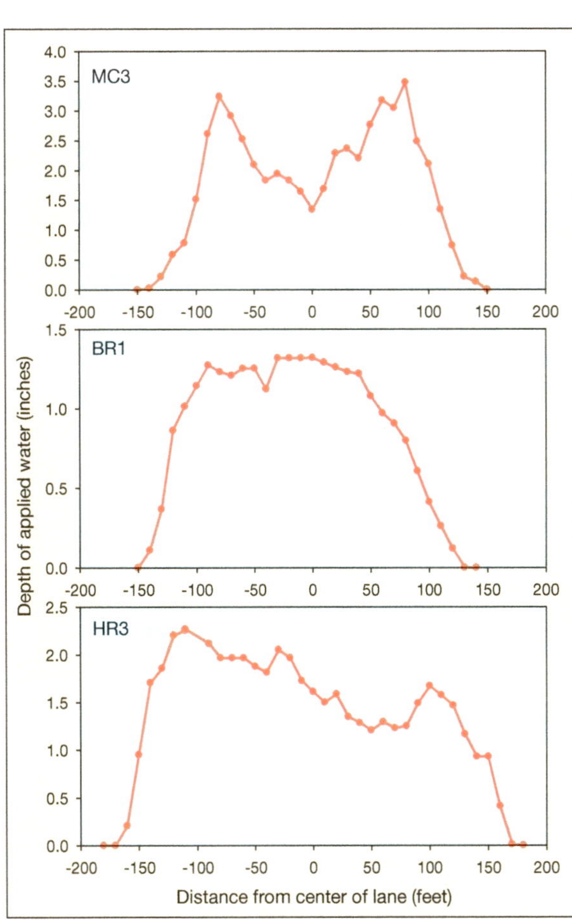

Figure 41. Nonoverlapped water application patterns for three traveling big gun sprinkle irrigation systems, designated MC3, BR1, and HR3. (We acknowledge the efforts of Ken Anderson, former Farm Advisor, Humboldt County, in obtaining these data.)

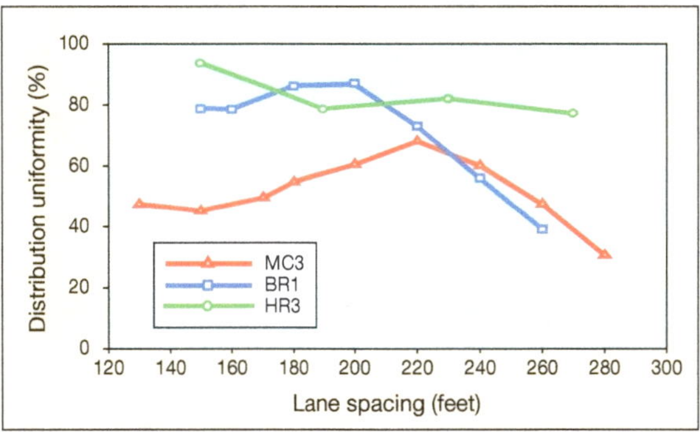

Figure 42. Distribution uniformity for various lane spacings for three different traveling sprinkle irrigation systems (designated as MC3, BR1, and HR3).

Evaporation and Wind Drift Considerations

Losses from Evaporation and Wind Drift

Water losses from spray evaporation and wind drift reduce the amount of water reaching the ground or plant canopy. Evaporation of water from sprinkle droplets occurs when droplets travel from the sprinkler nozzle to the ground or plant surface. Wind drift occurs when the droplets are blown away from the area intended to be irrigated. Evaporation and wind drift depend on climatic factors such as solar radiation, air temperature, humidity, and wind speed, as well as sprinkler characteristics such as pressure, nozzle diameter, riser height, and trajectory of the sprinkler jet.

Evaporation Losses

How much water is lost to spray evaporation and drift? Some studies have reported spray evaporation losses as high as 40 to 50 percent. However, these results may reflect the experimental procedure used to measure the losses, such as using a single sprinkler located over a bare surface. Evaporation losses were calculated as the difference between the volume of water discharged from the sprinkler head and the volume collected in catch cans. Factors contributing to errors in the loss calculation include higher evaporation rates of a single sprinkler head than would occur for multiple sprinkler heads such as in a field under irrigation, evaporation from the catch cans during the test, and wind drift off the measurement site. Sometimes the reported evaporation loss is a combination of evaporation and wind drift losses.

A University of Nebraska study simulated field conditions of a plant canopy and multiple sprinkler heads (Yazar 1984). A wide range of wind speeds, air temperatures, and relative humidity was used. Spray evaporation was found to be 5 percent or less for wind speeds up to 15 miles per hour and for relative humidity greater than about 50 percent. For low humidity conditions, the average evaporation loss was about 14 percent for wind speeds between 10 and 15 miles per hour. Losses due to drift only were less than 6 percent for wind speeds up to 10 miles per hour. These results suggest that losses due to evaporation and drift generally are less than 10 percent. Higher losses could occur for wind speeds greater than about 10 miles per hour and for low relative humidity.

Nomograph

Figure 43 contains a nomograph for estimating spray evaporation. Calculated evaporation using this nomograph and the climate data of the Nebraska study agreed well with the measured evaporation of the Nebraska study. Data needed for this nomograph are relative humidity, air temperature, and wind speed (all available from CIMIS weather stations), nozzle diameter expressed in 64ths of an inch, and sprinkler pressure. The procedure for using the nomograph is as follows:

Step 1. Draw a straight line from the appropriate relative humidity value (1) to the approximate temperature (2) until it intersects the vapor pressure deficit axis (3).

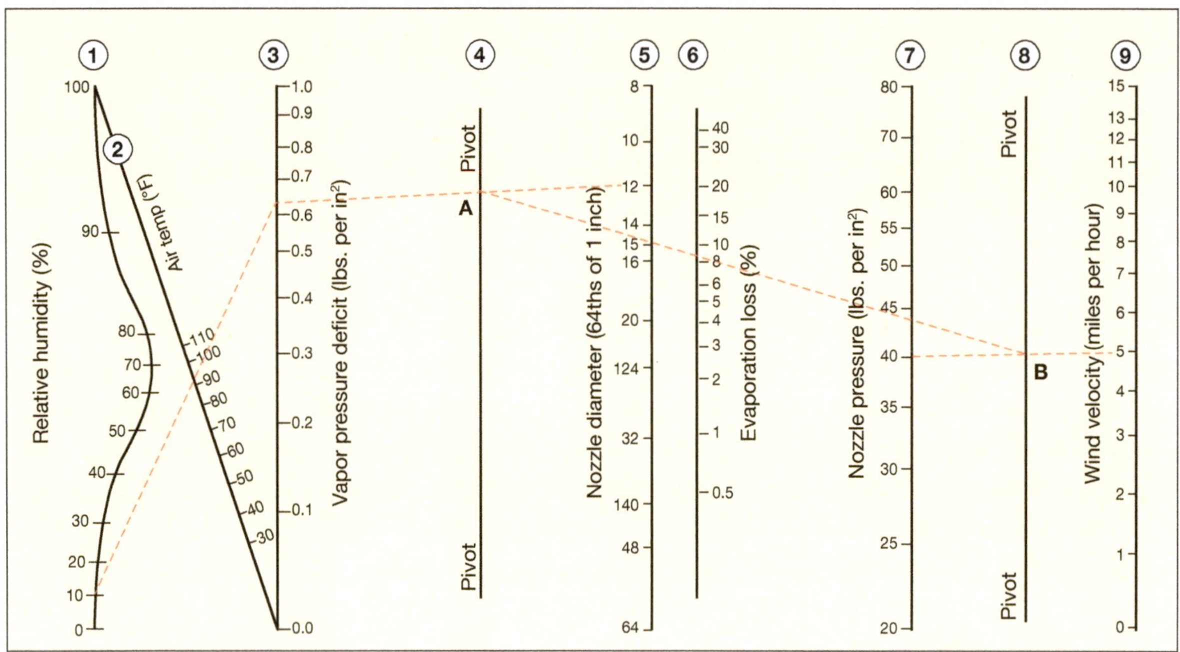

Figure 43. Nomograph for determining evaporation of water during the travel time of sprinkler droplets. *Source:* Frost and Schwalen 1955.

Step 2. Next draw a line from the intersection point on the vapor pressure deficit axis (3) to the nozzle diameter for your sprinkler system (5). A pivot point exists where this line crosses the pivot axis line (4).

Step 3. Draw a line from the wind speed axis (9) to the nozzle pressure axis (7). A second pivot point exists where this line crosses the pivot axis (8).

Step 4. Now draw a line between the two pivot points. The evaporation loss is where this line intercepts the percentage evaporation loss axis (6).

Evaporation Losses in California

The nomograph (see fig. 43) was used to determine the average daily evaporation losses at different locations in California using CIMIS climatic data (table 18). Results showed the average daily evaporation losses during a 24-hour period to range from 7 to 12 percent of the applied water.

Night and Day Sprinkle Irrigation

Is sprinkle irrigation at night more efficient than during the day? Will night irrigation reduce spray evaporation and wind drift as well as increasing yield?

Night and Day Climate Characteristics

During the night, no solar radiation occurs, and thus air temperatures are at a minimum and humidity is at a maximum, as shown for various locations in California (see table 18). At the same time, wind speed usually decreases during the night. After sunrise, air temperature increases to maximum values near midafternoon, humidity decreases to minimum values near midafternoon, and wind speed typically increases to maximum values in the late evening, depending on weather characteristics of a particular day. The effect of decreased air temperature, decreased wind speed, and increased humidity at night reduces the evaporation of water. Wind drift is also decreased due to the lower wind speeds, which can increase the uniformity of applied water at night compared with daytime.

Night and Day Evaporation and Wind Drift

A comparison of day and night sprinkle irrigation showed evaporation and drift losses of 16.5 percent and 7.4 percent for day and night irrigations, respectively, for one study (Playán et al. 2005), and respective losses of 21 and 11 percent in a second study (Cavero et al. 2008). For both studies, lower wind speeds, lower air temperatures, and higher humidity occurred during the night than during the day. Differences in wind speed appeared the most significant factor affecting evaporation and drift.

An estimate of day (0600 to 1800 hours) and night (1800 to 0600 hours) evaporation was made for July 15, 2008, using CIMIS data and the nomograph of Frost and Schwalen (see fig. 43). Results, in table 19, show higher evaporation during the day than at night, but the differences are site

Table 18. Climate characteristics and evaporation at different times of day for various locations throughout California

Time of day	RH (%)	T (°F)	WS (mph)	E (%)
		Tulelake		
0–3	70	59	2.6	4
3–6	88	54	1.7	3
6–9	65	65	2.0	4
9–12	48	76	3.7	7
12–15	34	81	4.6	10
15–18	36	81	4.9	10
18–21	50	72	6.0	8
21–24	67	64	3.6	5
Average				7
		Sacramento Valley (Colusa)		
0–3	63	66	7.3	7
3–6	75	62	4.6	5
6–9	67	67	5.0	6
9–12	53	76	4.4	7
12–15	42	85	5.7	12
15–18	42	88	6.0	13
18–21	48	81	6.4	12
21–24	60	70	7.0	8
Average				9
		Sacramento Valley (Davis)		
0–3	69	61	6.9	6
3–6	75	59	5.6	5
6–9	66	65	6.2	6
9–12	53	74	5.1	8
12–15	42	82	6.2	10
15–18	38	84	9.2	15
18–21	49	72	9.2	11
21–24	61	63	7.2	8
Average				9

Time of day	RH (%)	T (°F)	WS (mph)	E (%)
		Salinas Valley (Arroyo Seco)		
0–3	90	59	5.2	5
3–6	89	59	4.6	4
6–9	86	60	5.0	5
9–12	70	69	5.9	6
12–15	59	75	11.8	15
15–18	68	69	14.3	10
18–21	85	61	10.4	7
21–24	89	58	7.0	5
Average				7
		San Joaquin Valley (Five Points)		
0–3	60	72	7.9	8
3–6	69	67	5.1	6
6–9	68	69	7.5	7
9–12	57	79	7.7	9
12–15	44	88	7.0	11
15–18	39	91	7.7	15
18–21	48	84	8.7	13
21–24	58	77	7.2	10
Average				10
		Imperial Valley (Calipatria)		
0–3	59	78	1.7	5
3–6	67	75	2.8	5
6–9	43	91	2.6	9
9–12	20	103	3.1	15
12–15	20	109	4.7	18
15–18	25	110	3.7	17
18–21	41	97	5.3	14
21–24	57	88	6.1	10
Average				12

Note: RH = relative humidity, expressed in percent; T = temperature, expressed in degrees Fahrenheit; WS = wind speed, expressed in miles per hour; and E = average daily evaporation losses, expressed in percentage of applied water. The time of day is based on the 24-hour time.

specific. A small difference occurred for the southern San Joaquin Valley because on that date wind speeds did not decrease much during the night compared with daytime speeds, whereas, for the Imperial Valley, night wind speeds were much lower than those during the day, resulting in much less evaporation at night than during the day (see table 19).

Table 19. Day and night evaporation for various locations in California on July 15, 2008

	Evaporation (%)	
Site	**Night**	**Day**
Tulelake	6	8
Sacramento Valley (Colusa)	8	10
Sacramento Valley (Davis)	8	10
Salinas Valley (Arroyo Seco)	6	9
southern San Joaquin Valley (Five Points)	9	11
Imperial Valley (Calipatria)	9	15

Effect of Night and Day Irrigation on Yield

Few studies exist on the effect of night and day sprinkle irrigation on crop yield. In one study, however, night irrigation increased maize yield by 10 percent compared with the yield produced using day irrigation (Cavero et al. 2008). This yield increase was believed to be due to an increase in the uniformity of the applied water and a reduction in the evaporation and drift losses. Day sprinkle irrigation reduced the uniformity of applied water by 5 to 7 percent. Average evaporation and drift losses were 11 percent for night irrigation and 21 percent for day irrigation.

Considerations

Sprinkle irrigation at night can reduce wind drift and evaporation losses and increase the uniformity of applied water, both of which can potentially increase yield. Thus, night irrigation is recommended.

However, practical considerations prevent many irrigators from irrigating at night only. The main consideration is the irrigation system flow rate. Flow rates of many sprinkle irrigation systems are such that near-continuous irrigation is needed to supply sufficient water to the crop during the summer, thus preventing night irrigation only. Also, more sprinkler laterals are needed to cover the entire field within a given irrigation interval for night-only irrigation, thus greatly increasing the capital costs of the irrigation system.

Energy Considerations

Pump Selection

Selecting the proper pump involves choosing one that provides the desired capacity or flow rate and total head while operating at close to maximum efficiency. Operating near maximum efficiency uses the least possible horsepower and energy, and it incurs the lowest possible demand charges. Detailed information on irrigation pumping plants can be found in *Irrigation Pumping Plants,* ANR Publication 3377 (Hanson 2000), which can be ordered from the ANR Catalog Web site, http://anrcatalog.ucdavis.edu.

Selecting a Pump

The irrigation pump must supply the flow rate and total head needed for satisfying crop water requirements. The flow rate of the pump must meet the crop water needs during the period of maximum evapotranspiration (see the section "Determining the Flow Rate").

Total head of a deep well turbine pipe consists of the pumping lift and the discharge pressure head. The pumping lift is the elevation difference between the pumping water level in the well and the pump discharge point. The pump discharge pressure is the pressure at the pump. The total head in feet is the sum of the pumping lift and discharge pressure head (pressure in psi × 2.31 feet per psi). The total head of a booster pump (used for supplying pressure) is simply the pressure difference between pump intake and pump discharge in psi, multiplied by 2.31 to obtain the pressure head.

Pump Performance Curves

Pump performance curves are used to select irrigation pumps. Performance curves describe the performance characteristics of deep well turbines and centrifugal or booster pumps by showing the relationship between total head and flow rate (fig. 44), pump efficiency and flow rate (fig. 45),

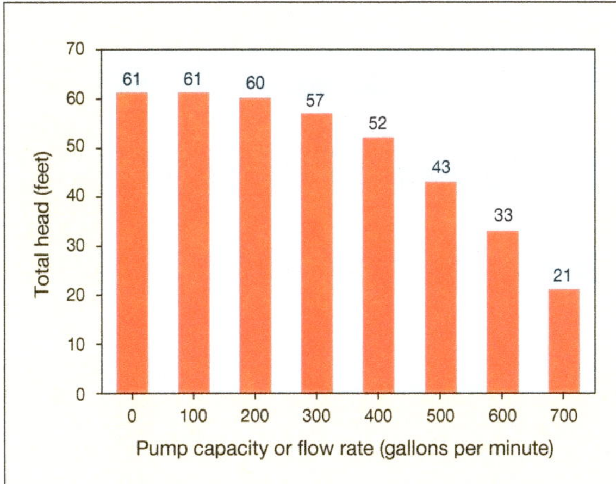

Figure 44. An example of a performance curve showing the relationship between total head and pump capacity or flow rate for a specific pump. Different pumps have different performance curves.

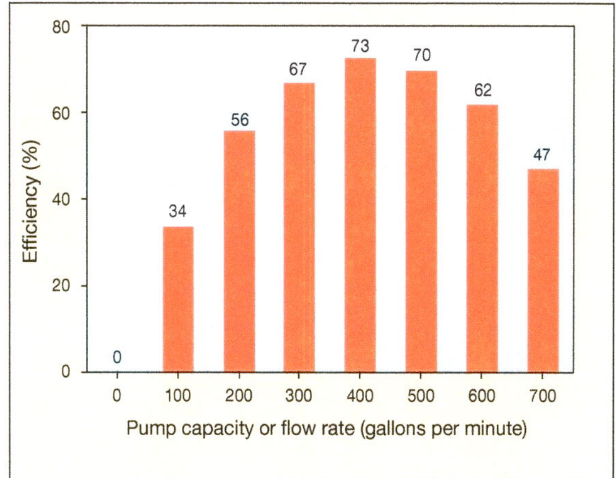

Figure 45. An example of a performance curve showing the relationship between pump efficiency and pump capacity or flow rate of a specific pump. Different pumps have different performance curves.

and brake horsepower and flow rate (fig. 46). Total head decreases as flow rate increases, pump efficiency increases to a maximum and then decreases as flow rate increases, and brake horsepower increases as flow rate increases.

Deep well turbine pumps normally consist of a series of stages, where one stage is a pump housing and an impeller. Pump performance curves are normally developed for a one-stage pump, that is, one pump housing containing one impeller. For a multiple-stage pump, the total head and brake horsepower of a single stage at a given flow rate are multiplied by the number of stages to obtain the performance curves of a multiple-stage pump.

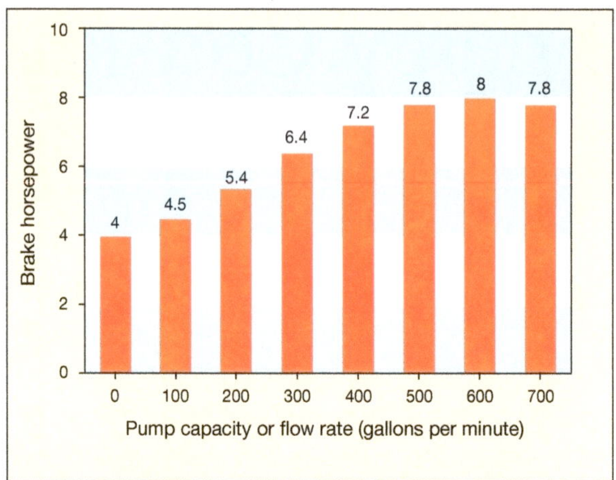

Figure 46. An example of a performance curve showing the relationship between pump brake horsepower demand and pump capacity or flow rate for a specific pump. Different pumps have different performance curves.

Selection Steps

The performance curves can be used to select an appropriate pump for the desired operating characteristics. Steps for choosing a pump are as follows:

Step 1. Estimate the flow rate and total head needed by the sprinkle irrigation system.

Step 2. Consult catalogs of manufacturers' pump performance curves to find a pump that will provide the needed flow rate and total head at close to maximum efficiency.

Step 3. Use the pump performance curve to determine the horsepower requirement of the pump.

Step 4. Select the appropriate electric motor or engine.

Factors Affecting Pumping Plant Performance

Major factors affecting pumping plant performance include the following:

- *Pump wear.* Sand in the well water causes pump wear, which reduces the total head, flow rate, and pumping plant efficiency.
- *Declining groundwater levels.* This decreases the pump flow rate. Surging can occur if the water level in the well reaches the level of the pump intake.
- *Expanding the irrigation system.* This reduces pressure and increases the flow rate of a pump. The flow rate of an expanded system is likely to be too small to provide adequate water to the expanded system.
- *Mismatched pump.* A mismatched pump is one that is operating properly but not at or near its maximum efficiency. This can be caused by declining groundwater levels or an expanded irrigation system.

- *Well clogging*. This restricts the flow of water into the well, thus increasing the pumping lift and reducing the flow rate.
- *Air or gas in water.* Air in the water reduces the flow rate. Sources of air include leaks in the suction side of a booster pump, cascading water in the well, or naturally occurring gas in the groundwater.
- *Cavitation*. This occurs when an excessive vacuum develops in the pump intake just prior to the impeller. Cavitation results in small air bubbles being formed and carried into the impeller, which damages the impeller and changes the pump's performance. A rattling noise that sounds like gravel in the pump indicates cavitation.

Pump Performance Tests

Pump tests can determine the effect of changes over time on pump performance. Pump tests are conducted by utility companies, pump dealers, and consultants. The test may provide any or all of the following information:

- pumping lift
- static or standing water level (i.e., water level in the well prior to pumping)
- pumping water level
- drawdown (i.e., elevation difference between the pump water level and the static water level)
- discharge pressure
- discharge pressure head
- pump flow rate
- well yield (i.e., flow rate in gallons per minute ÷ drawdown). (A well yield decreasing over time may indicate plugging in the well.)
- acre-feet pumped per 24 hours
- input horsepower or kilowatts of engine or motor
- percentage motor overload
- kilowatt-hours per acre-foot of pumped water
- overall pumping plant efficiency. (The overall efficiency includes the efficiency of the electric motor or engine and is frequently referred to as the "wire to water" efficiency for electric motors. Note that the efficiencies shown in the pump performance curve is the pump efficiency only, not the pumping plant efficiency.)

Results of the pumping plant test can be interpreted using table 20 for pumps driven by an electric motor.

Table 20. Recommended corrective action to improve pumping plant efficiency at various efficiency levels

Efficiency rate	Corrective action
> 60%	no corrective action
55% to 60%	consider an impeller adjustment
50% to 55%	consider an impeller adjustment; consider repairing or replacing pump if impeller adjustment has no effect
< 50%	consider repairing or replacing pump

Variable Speed Drives for Irrigation Pumping Plants

Electric motors used for irrigation pumping plants operate at a constant rpm (revolutions per minute). However, situations occur when this is not optimal for the most efficient operation. The same pump may be used to separately irrigate fields of different sizes or irrigate irregularly shaped fields with lateral lengths that change during the field irrigation. A pump may be selected that has sufficient pressure and flow rate to irrigate the largest acreage, but the pump's output may be excessive for the smaller irrigated acreage, which generally requires smaller flow rates than those needed for the largest area. To prevent excessive pressure, a partially closed valve at the pump discharge is commonly used to reduce pressure. This practice results in a substantial energy loss across the valve and a low pumping plant efficiency.

A variable speed drive can overcome the problems associated with using the same pump for different irrigated acreages. A variable speed drive is an electronic device connected to the power panel that allows the rpm of the motor to be reduced by varying the frequency of the power into the motor. Decreasing the frequency reduces the power to the motor and, as a result, the electric motor's rpm decreases. This decrease also decreases the pump flow rate, total head, and brake horsepower. Because of the relationship between pump rpm and pump horsepower demand, a small decrease in rpm greatly decreases the brake horsepower demand of the pump, resulting in significant energy savings.

Example

A centrifugal pump was used to irrigate two fields separately of 80 acres and 50 acres with water supplied from a reservoir. A pump test was first conducted while irrigating the 80-acre field with no restriction in the pump discharge. A second test was conducted while irrigating the 50-acre field with a partially closed valve (normal grower practice) to reduce the pressure. A third test was conducted using the variable speed drive to reproduce the output of the second pump test without the partially closed valve.

The test while irrigating the 80-acre field showed an overall efficiency of 40 percent for a flow rate of 1,100 gallons per minute and a discharge pressure of 80 psi (table 21). Input horsepower was 128. The overall efficiency is considered to be low, indicating room for improvement.

Pressure, flow rate, input horsepower, and overall pumping plant efficiency measured during a second pump test were 64 psi, 600 gallons per minute, 90 horsepower, and 22 percent, respectively (see table 21). The third test showed a pressure and flow rate of 60 psi and 700 gallons per minute, respectively (see table 21). The variable speed drive reduced the pump rpm from 1,770 (normal rpm) to 1,345, which in turn reduced the input horsepower to 55, a 39 percent reduction compared with the second test's input horsepower. Note that the pumping plant efficiency of the variable speed operation was nearly the same as that of the 80-acre operation, but it was much higher than the partially closed valve condition.

Table 21. Pump test results using the same centrifugal pump to irrigate different-sized fields

	Test 1—valve fully open, normal pump rpm (80 acres)	Test 2—partially closed valve, normal pump rpm (50 acres)	Test 3—variable speed drive, valve fully open (50 acres)
rpm	1,770	1,770	1,345
pressure (psi)	80	64	60
pump capacity (gpm)	1,100	600	700
input horsepower	128	90	55
overall efficiency (%)	40	22	44

Thus, using a variable speed drive would reduce the energy costs by 39 percent, assuming the same irrigation set times for the partially closed valve condition and the variable speed condition.

Reducing Energy Use of Sprinkle Irrigation Systems

The measurement of energy use is based on kilowatt-hours or horsepower-hours consumed. The term *kilowatt-hours* is normally used for electric pumping plants and *horsepower-hours* for engines. A kilowatt (kw) is directly related to horsepower (1 kw = 0.746 horsepower) and is the rate of energy use per unit of time. Hours refer to the operating time of the pumping plant. Thus, reducing energy use and, hopefully, energy costs, requires reducing the kilowatt or horsepower demand, reducing the operating time, or a combination of the two. Understanding the factors affecting energy use is important because some measures promoted as energy-saving measures, while increasing efficiency, can actually increase energy use.

Measures for potentially reducing the energy use of sprinkle irrigation systems include the following:

- *Reduce the operating pressure of the irrigation system.* This involves replacing the sprinkler nozzles with larger ones that have the same discharge rate at the lower pressure as the original nozzles at the higher pressure. It also involves modifying the pump by trimming the impellers to reduce their diameter. Trimming the impellers reduces horsepower demand of the pump and also reduces total head and flow rate. The pressure and flow rate of the trimmed impeller must be high enough for proper sprinkler performance.

- Simply replacing the existing nozzles with larger nozzles without modifying the pump impeller decreases the pressure, but it also increases the flow rate, thus potentially increases the horsepower demand of the pump because of the relationship between pump brake horsepower and flow rate. For this condition, a potential for energy savings can occur only if the irrigation time is decreased.

- *Adjust or repair the pump.* The efficiency of pumps using semi-open impellers may be improved by an impeller adjustment. This involves adjusting the clearance between the pump housing and the bottom of the impeller vanes by slightly lowering the impellers using the adjustment nut located on the top of the electric motor. Repairing the pump involves removing the pump from the well and rebuilding or replacing worn impellers.

  While this measure can increase the overall pumping plant efficiency, it can also increase the energy use. Table 22 shows examples of the effect of impeller adjustments and repairs on pumping plant performance. While both measures (i.e., making impeller adjustments and repairing worn impellers) increase flow rate, total head, and pumping plant efficiency, the input horsepower also increases. Thus, operating the pump for the same time after the adjustment or repair as before increases energy costs because of the increased horsepower demand. Energy savings will be realized only by reducing the operating time.

Table 22. Total head, flow rate, overall pumping plant efficiency, and input horsepower before and after impeller adjustment or pump repair

Adjustment or repair		Flow rate (gpm)	Total head (feet)	Overall efficiency (%)	Input horsepower
Impeller adjustment					
pump 1	before	605	148	54	42
	after	910	152	71	49
pump 2	before	708	181	59	55
	after	789	206	63	65
pump 3	before	432	302	54	61
	after	539	323	65	67
pump 4	before	616	488	57	133
	after	796	489	68	144
Pump repair					
pump 5	before	878	227	46	109
	after	1,348	271	66	140
pump 6	before	989	75	32	58
	after	3,100	100	72	109
pump 7	before	1,090	329	45	201
	after	1,618	307	61	206
pump 8	before	664	180	41	74
	after	1,335	185	68	92

- Use appropriate irrigation set times needed to supply the desired amount of water. Irrigation set times that are too long increase energy use. Evaluate the irrigation set time with soil moisture sensors or by estimating crop ET between irrigations and calculating the set time required to apply the necessary amount of water.

- Replace a mismatched pump. A mismatched pump operates properly but not at its point of maximum efficiency. Replacing a mismatched pump with a pump that provides the same output at or near its point of maximum efficiency will reduce the kilowatt or horsepower demand.

- Use a variable speed drive.

- Do not operate with a partially closed valve in the pipeline.

- Use an appropriate electric power rate plan.

- Consider using larger pipe diameters for sprinkler laterals when replacing the irrigation system to reduce pressure losses due to friction.

- Consider well maintenance to improve well efficiency. Encrustation in the well can reduce the openings for water to flow into the well, thus increasing the pumping lift and decreasing the flow rate.

- Maintain the sprinkle irrigation system. Proper maintenance may have little effect on energy use, but it will promote maximum return.

Appendix A

Table A-1. Historical reference crop evapotranspiration (ET_0) in inches per day

		Shafter	Five Points	Parlier	Davis	Nicolaus	Durham	McArthur	Brawley
Jan.	1–15	0.03	0.04	0.03	0.03	0.03	0.03	0.02	0.07
	16–31	0.05	0.05	0.04	0.05	0.04	0.05	0.03	0.09
Feb.	1–15	0.07	0.06	0.06	0.06	0.06	0.06	0.04	0.10
	16–30	0.09	0.09	0.08	0.09	0.09	0.09	0.07	0.13
Mar.	1–15	0.11	0.11	0.10	0.09	0.09	0.09	0.08	0.16
	16–31	0.14	0.15	0.13	0.14	0.12	0.12	0.11	0.19
Apr.	1–15	0.19	0.20	0.17	0.18	0.15	0.16	0.14	0.22
	16–30	0.20	0.22	0.19	0.20	0.18	0.17	0.14	0.25
May	1–15	0.24	0.26	0.22	0.23	0.21	0.21	0.18	0.28
	16–31	0.26	0.27	0.24	0.24	0.21	0.22	0.19	0.29
June	1–15	0.27	0.29	0.26	0.28	0.24	0.25	0.22	0.31
	16–30	0.28	0.30	0.27	0.29	0.26	0.26	0.25	0.32
July	1–15	0.28	0.30	0.27	0.29	0.26	0.27	0.27	0.31
	16–31	0.26	0.28	0.25	0.27	0.25	0.25	0.25	0.29
Aug.	1–15	0.25	0.28	0.24	0.26	0.24	0.24	0.25	0.29
	16–31	0.23	0.25	0.22	0.24	0.21	0.21	0.22	0.28
Sept.	1–15	0.21	0.23	0.19	0.21	0.19	0.19	0.18	0.26
	16–30	0.18	0.20	0.15	0.18	0.16	0.16	0.14	0.22
Oct.	1–15	0.16	0.17	0.13	0.16	0.13	0.14	0.12	0.19
	16–31	0.12	0.13	0.09	0.12	0.09	0.10	0.08	0.15
Nov.	1–15	0.08	0.10	0.07	0.09	0.07	0.07	0.05	0.12
	16–30	0.06	0.07	0.04	0.06	0.05	0.05	0.03	0.10
Dec.	1–15	0.05	0.05	0.03	0.05	0.03	0.04	0.02	0.07
	16–31	0.03	0.03	0.02	0.04	0.04	0.03	0.02	0.07

Table A-2. Typical root depths containing approximately 80 percent of the feeder roots for various crops in a deep, uniform, well-drained soil profile

Crop	Root depth (feet)
alfalfa	4.0–6.0
almond	2.0–4.0
apple	2.5–4.0
apricot	2.0–4.5
artichoke	2.0–3.0
asparagus	6.0
avocado	2.0–3.0
banana	1.0–2.0
barley	3.0–3.5
bean	
dry	1.5–2.0
green	1.5–2.0
lima	3.0–5.0
beet	
sugar	1.5–2.5
table	1.0–1.5
berries	3.0–5.0
broccoli	2.0
Brussels sprouts	2.0
cabbage	2.0
cantaloupe	2.0–4.0
carrot	1.5–2.0
cauliflower	2.0
celery	2.0
chard	2.0–3.0
cherry	2.5–4.0
citrus	2.0–4.0
coffee	3.0–5.0
corn	
grain and silage	2.0–3.0
sweet	1.5–2.0
cotton	2.0–6.0
cucumber	1.5–2.0
eggplant	2.5
fig	3.0
flax	2.0–3.0

Crop	Root depth (feet)
grape	1.5–3.0
lettuce	0.5–1.5
lucerne	4.0–6.0
oat	2.0–2.5
olive	2.0–4.0
onion	0.5–1.0
parsnip	2.0–3.0
passion fruit	1.0–1.5
pasture (annual)	1.0–2.5
pasture (perennial)	1.0–2.5
pea	1.5–2.0
peach	2.0–4.0
pear	2.0–4.0
pepper	2.0–3.0
plum	2.5–4.0
potato	
Irish	2.0–3.0
sweet	2.0–3.0
pumpkin	3.0–4.0
radish	1.0
safflower	3.0–5.0
sorghum	
grain and sweet	2.0–3.0
silage	3.0–4.0
soybean	2.0–2.5
spinach	1.5–2.0
squash	2.0–3.0
strawberry	1.0–1.5
sudangrass	3.0–4.0
sugarcane	1.5–3.5
tobacco	2.0–4.0
tomato	2.0–4.0
turnip (white)	1.5–2.5
walnut	5.5–8.0
watermelon	2.0–3.0
wheat	2.5–3.5

Source: USDA 1983.
Note: There may be many exceptions to these guidelines because of soil types, restrictive layers in the soil profile, and irrigation practices.

Appendix B

Crop Coefficients of Various Row Crops

Crop coefficients (K_c) and reference crop ET (ET_o) frequently are used to calculate the evapotranspiration (ET) of a crop where:

$$ET = K_C \times ET_O$$

Crop coefficients vary with crop type and stage of growth. Crop coefficients are small during the initial growth stage (fig. B-1). However, during sprinkle irrigation or rainfall, coefficients can be nearly equal to 1.0, due primarily to high evaporation rates from the soil surface. After the irrigation, crop coefficients rapidly decrease as the soil surface dries. An average crop coefficient for this early growth stage should account for sprinkle irrigation and rainfall. Crop coefficients increase rapidly during the canopy development growth stage of the crop. Maximum values occur during the midseason growth stage, but crop coefficients can decrease during the late-season growth stage. Some crops such as lettuce, however, are harvested before the late-season growth stage occurs. It is generally assumed that the canopy cover for initial growth stages is 10 percent or less, the start of the canopy development stage occurs at about 10 percent canopy cover, and the canopy cover is 70 to 80 percent at the start of the midseason stage.

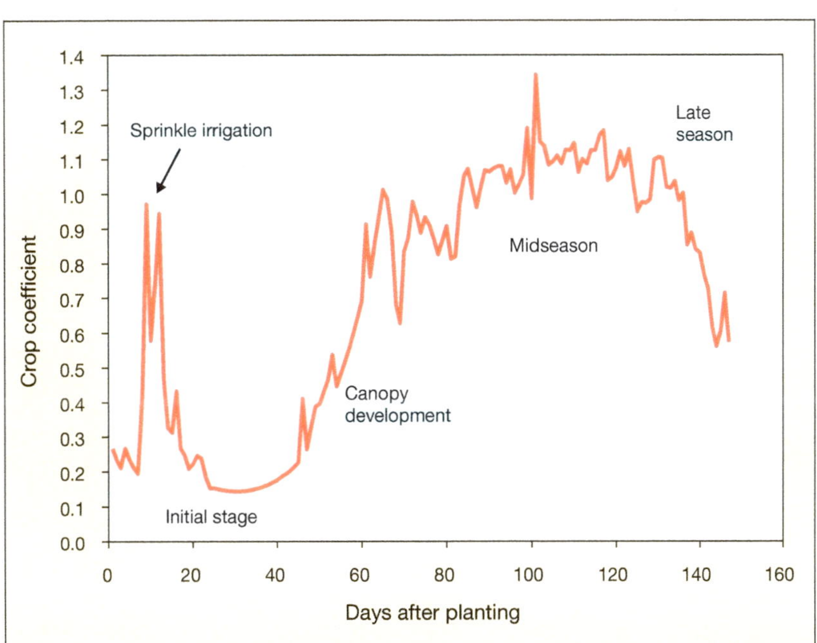

Figure B-1. Crop coefficients of processing tomatoes over time for the various stages of growth.

Crop coefficient values are found in several publications (Snyder et al. 1989a; Allen et al. 1998). They are frequently described on a time basis, such as calendar day, day of the year, or days after planting or emergence. These values are considered to be universal, but there is some uncertainty because crop coefficients developed in one area may differ from those developed elsewhere due to different conditions. However, in many cases these may be the only values available. When possible, use locally developed crop coefficients. Otherwise, crop coefficients may require some adjustment based on soil moisture monitoring, observations of canopy development, or irrigator experience. Also, crop coefficients for the canopy development growth stage must be estimated because these coefficients are rarely provided.

Crop coefficients expressed on a canopy cover basis overcome the problems that occur using crop coefficients expressed on a time basis; however, these data are limited. The canopy cover is defined as the bed area shaded by the canopy at midday. The canopy cover is determined by first measuring the width of the plant canopy (by eyeballing it) and then dividing that value by the bed spacing or

$$\text{Canopy cover (\%)} = 100 \times \text{canopy width} \div \text{bed spacing}$$

This measurement approach can be used any time of the day since the actual canopy cover is being estimated instead of the shaded area. Estimating the canopy width is relatively easy for crops such as tomatoes, peppers, lettuce, and cotton, but may not be possible for onions and melons. These relationships are particularly useful for determining the crop coefficients of the canopy development growth stage.

Example

The canopy width of tomatoes was estimated to be 40 inches. The bed spacing is 60 inches. Thus, the canopy cover is 100 × 40 inches ÷ 60 inches = 50%. The crop coefficient, using figure B-2, is 0.8.

Crop coefficients expressed on the basis of heat units or growing degree days (GDD) also overcome the problems associated with expressing crop coefficients on a time basis. The heat unit method, however, requires research that correlates crop coefficients with the cumulative heat units, which has been done for cotton and tomatoes in California. Also the heat units accumulated during the crop season for a particular year must be calculated. This is generally not done in California. Relationships between crop coefficients and cumulative heat units are available for some vegetable crops from the University of Arizona Cooperative Extension.

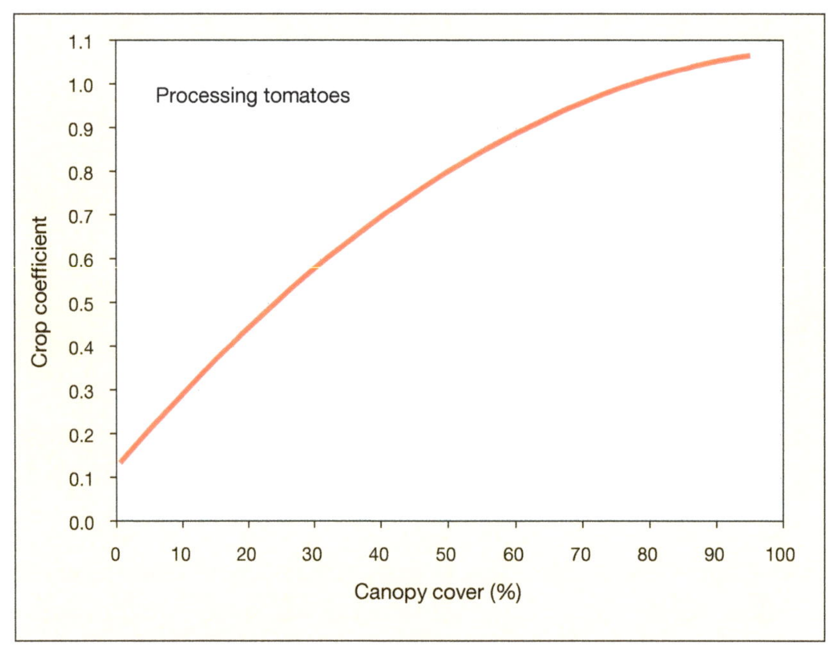

Figure B-2. Canopy cover-crop coefficient relationship for processing tomatoes. *Source:* Hanson and May 2006b.

It is recommended that the initial stage crop coefficients (when canopy coverage is smaller than 10 percent) not be used for irrigation water management, that is, determining when to irrigate. Irrigating when total ET between irrigations equals the allowable soil moisture depletion does not apply during this growth stage. Use of allowable soil moisture depletion as an irrigation criterion is based on the rooting depth of a mature crop. Rooting depths during the early growth stages are generally not known, and thus the amount of irrigation water and timing of irrigations should be based on irrigator experience for the local conditions. The initial growth stage crop coefficients and reference crop ET can be used to estimate ET during the early growth stages. This value can be compared to the amount of applied water to ensure that sufficient water is applied.

Crop coefficients of some row and field crops that are irrigated with sprinkle irrigation in California are presented in the following sections.

Alfalfa

Alfalfa crop coefficients are small just after harvest or cutting, and then they increase after the first postharvest irrigation to maximum values just before the next harvest. This cycle is repeated throughout the crop season. Because of the constantly changing crop coefficients, using actual coefficients can be difficult. Thus, an average crop coefficient for the crop season is normally used for alfalfa irrigation water management. Table B-1 lists average alfalfa crop coefficients for various areas of California.

Table B-1. Average crop coefficients for alfalfa in California

Location	Crop coefficient
Imperial Valley and Palo Verde Valley (low desert)	1.00
southern San Joaquin Valley	1.01
Sacramento Valley	0.95
Scott Valley (intermountain area of Northern California)	0.90
Tulelake (Klamath Basin)	0.98

Broccoli

Figure B-3 shows a relationship between canopy cover and crop coefficient for broccoli (Grattan et al. 1998).

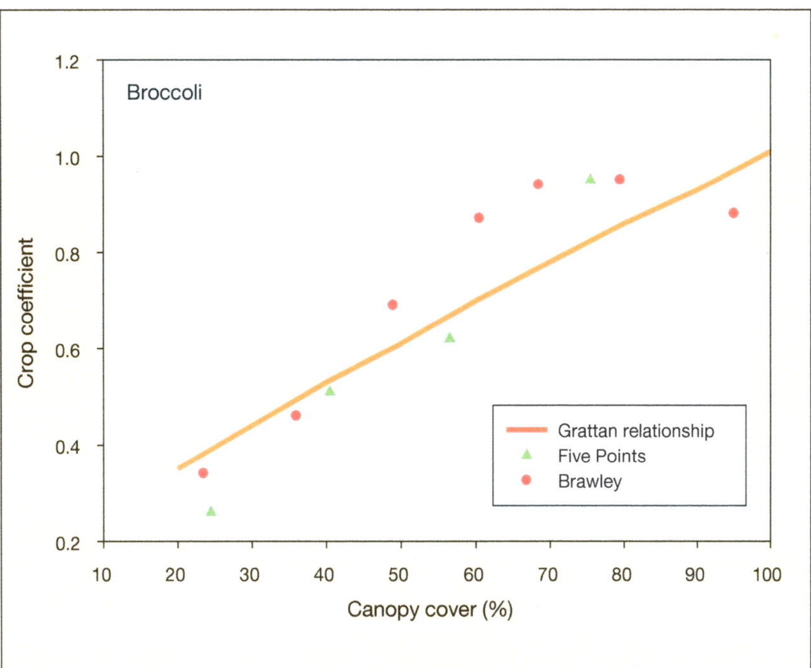

Figure B-3. Canopy cover-crop coefficient relationship for broccoli. *Source:* Grattan et al. 1998. Data points are from western Fresno County and the Imperial Valley and show a reasonable fit to the straight line. *Source:* R. Hutmacher, unpublished data.

Carrot

Carrot crop coefficients for the southern San Joaquin Valley are listed in table B-2 for planting dates of February 1 and September 1.

Cotton

A relationship between crop coefficient and canopy cover is shown in figure B-4 for two cotton varieties (R. Hutmacher. Unpublished data.) This relationship was developed on the west side of Fresno County.

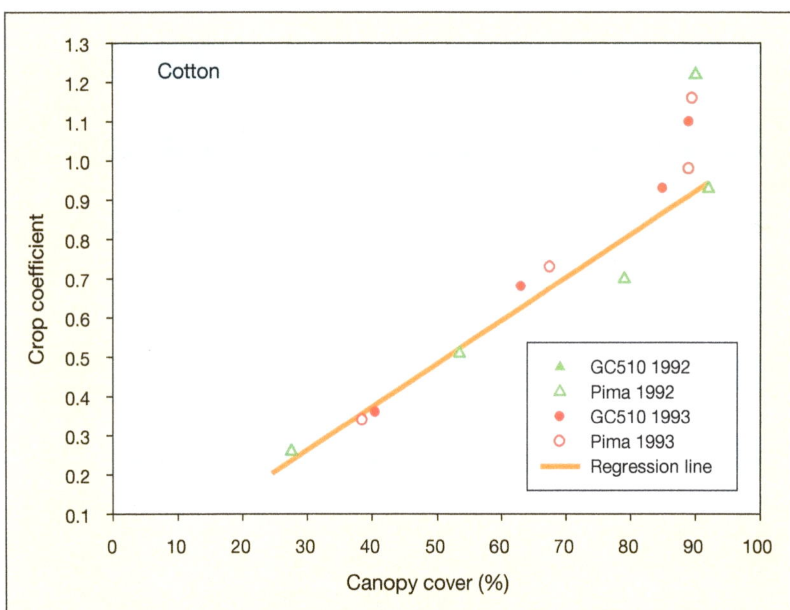

Figure B-4. Canopy cover-crop coefficient relationship for cotton. For canopy cover of about 90 percent or greater, the crop coefficient should be increased to about 1.1 to 1.2. *Source:* R. Hutmacher, unpublished data.

Table B-2. Carrot crop coefficients for planting dates of February 1 and September 1

	Crop coefficient	
	Planting date	
Date	**Feb. 1**	**Sept. 1**
Jan. 6		1.10
Jan. 13		1.05
Jan. 20		1.00
Jan. 27		0.90
Feb. 3	1.00	0.90
Feb. 10	1.00	
Feb. 17	0.09	
Feb. 24	0.13	
Mar. 3	0.13	
Mar. 10	0.18	
Mar. 17	0.27	
Mar. 24	0.35	
Mar. 31	0.55	
Apr. 7	0.82	
Apr. 14	0.99	
Apr. 21	1.07	
Apr. 28	1.06	
May 5	1.07	
May 12	1.11	
May 19	1.13	
May 26	1.13	
June 2	1.05	
June 9	1.05	
June 16	1.01	
June 23	0.98	
June 30	0.89	
July 7	0.90	
July 14		
July 21		
July 28		
Aug. 4		
Aug. 11		
Aug. 18		
Aug. 25		
Sept. 1		1.00
Sept. 8		1.00
Sept. 15		0.15
Sept. 22		0.30
Sept. 29		0.50
Oct. 6		0.70
Oct. 13		0.90
Oct. 20		1.00
Oct. 27		1.05
Nov. 3		1.10
Nov. 10		1.15
Nov. 17		1.15
Nov. 24		1.15
Dec. 1		1.15
Dec. 8		1.19
Dec. 15		1.19
Dec. 22		1.19
Dec. 29		1.10

Source: Estimated by B. Sanden, Kern County UC Cooperative Extension Web site, http://cekern.ucdavis.edu/Irrigation_Management/.

Lettuce

Published data on lettuce crop coefficients show values ranging from 0.17 to 0.70 for the initial stage and 0.83 to 1.02 for the midseason growth stage (Snyder et al. 1989a; Allen et al. 1998). The range of values is too large to be very useful to irrigators. Thus, it is recommended that the relationship between canopy cover and crop coefficient in figure B-5 be used since measuring the canopy cover of lettuce is fairly easy.

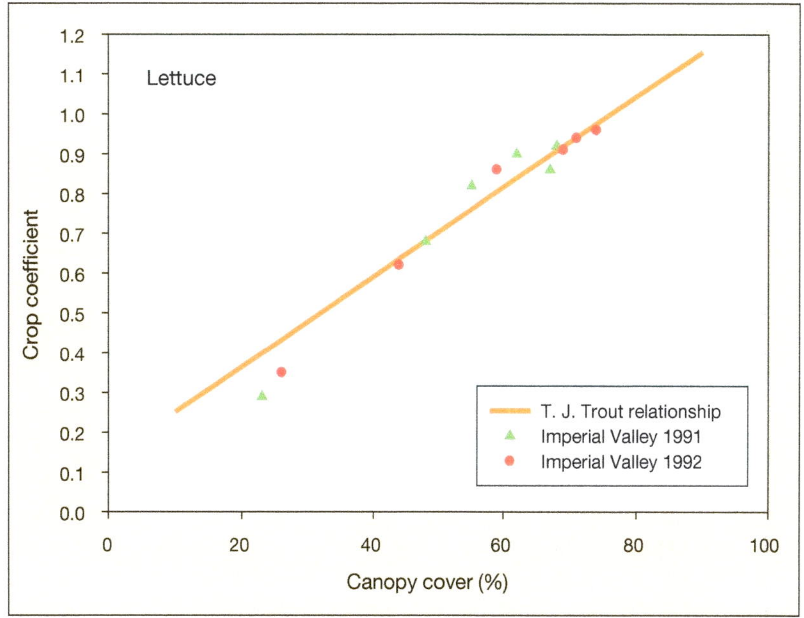

Figure B-5. Canopy cover-crop coefficient relationship for lettuce. *Source:* Trout and Johnson 2007. Data from the Imperial Valley (1991, 1992) show excellent agreement. *Source:* R. Hutmacher, unpublished data.

Onions

Figure B-6 shows a relationship between canopy cover and onion crop coefficient (López-Urrea et al. 2009). While this relationship was determined in Spain, these values of crop coefficients reflect published values (expressed on a time basis) for the San Joaquin Valley and the high desert area of Southern California. Because estimating canopy cover of onions is difficult, crop coefficients on a time of year basis are shown in table B-3 for several locations in California.

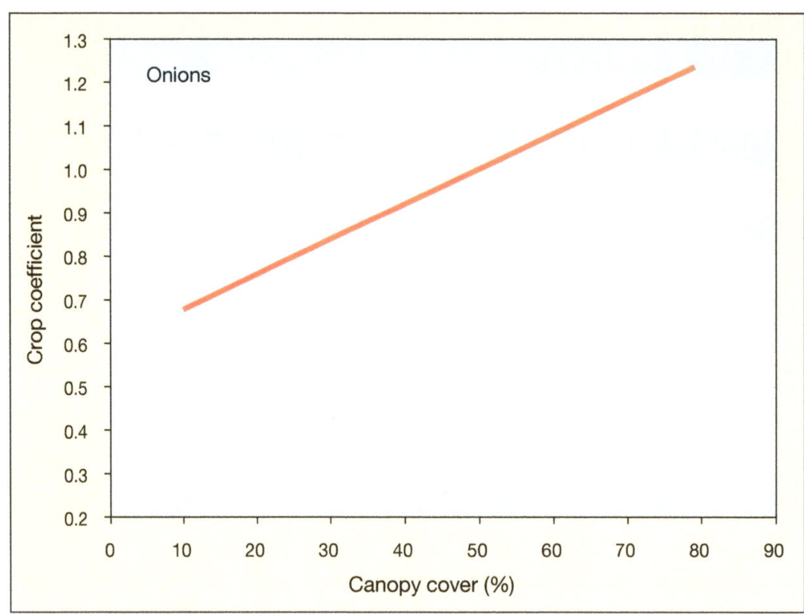

Figure B-6. Canopy cover-crop coefficient relationship for onions. *Source:* López-Urrea et al. 2009.

Table B-3. Onion crop coefficients for the high desert of Southern California, the San Joaquin Valley, and the Klamath Basin

High desert	
Date	Crop coefficient
Mar. 1	0.30
Mar. 15	0.30
Apr. 1	0.30
Apr. 15	0.53
May 1	0.83
May 15	1.14
June 1	1.14
June 15	1.14
July 1	1.04
July 15	0.92
Aug. 1	0.80
Aug. 15	0.68

San Joaquin Valley			
Date	Crop coefficient		
	Planting date		
	Mar. 1	Sept. 16	Nov. 16
Jan. 1		1.15	0.71
Jan. 15		1.15	0.98
Feb. 1		1.15	1.11
Feb. 15		1.15	1.11
Mar. 1	0.30	1.15	1.11
Mar. 15	0.30	1.12	1.11
Apr. 1	0.30	1.05	1.11
Apr. 15	0.53	0.97	1.11
May 1	0.83	0.90	1.11
May 15	1.14	0.82	1.11
June 1	1.14		1.11
June 15	1.14		1.06
July 1	1.04		0.86
July 15	0.92		0.64
Aug. 1	0.80		
Aug. 15	0.68		
Sept. 1			
Sept. 15		0.18	
Oct. 1		0.20	
Oct. 15		0.38	
Nov. 1		0.55	
Nov. 15		0.72	0.27
Dec. 1		0.88	0.27
Dec. 15		1.06	0.47

Klamath Basin	
Date	Crop coefficient
Apr. 14	0.11
Apr. 21	0.11
Apr. 30	0.11
May 7	0.14
May 14	0.23
May 21	0.33
May 31	0.46
June 7	0.56
June 14	0.65
June 21	0.74
June 30	0.87
July 7	0.96
July 14	1.06
July 21	1.20
July 31	1.20
Aug. 7	1.17
Aug. 14	1.09
Aug. 21	0.99
Aug. 31	0.86
Sept. 7	0.77
Sept. 14	0.69
Sept. 21	0.62
Sept. 25	0.57

Sources: Snyder et al. 1989a (high desert of Southern California and San Joaquin Valley); estimated by Harry Carlson, Farm Advisor (retired), Modoc County (Klamath Basin).

Peppers

A relationship between crop coefficient and canopy cover for peppers is shown in figure B-7. Data were developed on the west side of Fresno County.

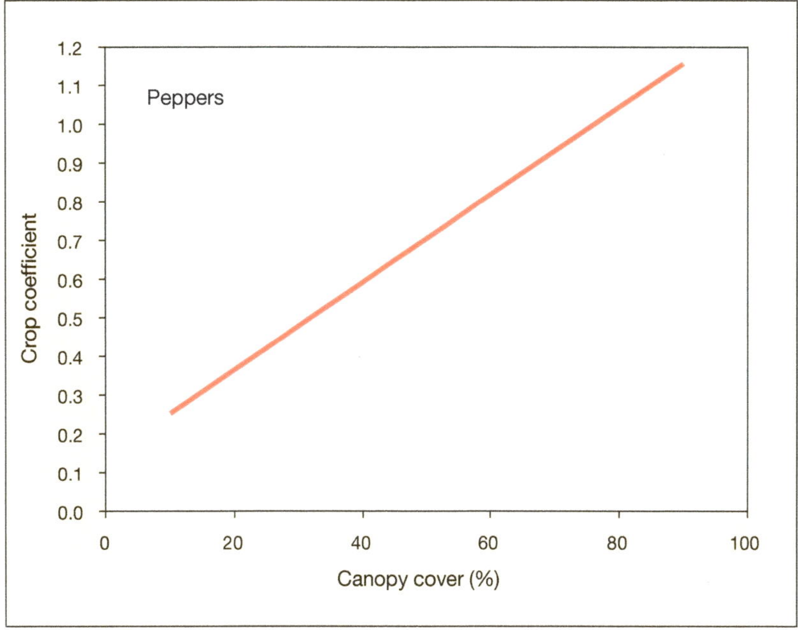

Figure B-7. Canopy cover-crop coefficient relationship for peppers. *Source:* Trout and Johnson 2007.

Potatoes

Potato crop coefficients expressed on a time of year basis for the southern San Joaquin Valley and the Klamath Basin are shown in table B-4.

Table B-4. Potato crop coefficients for the southern San Joaquin Valley and the Klamath Basin

Southern San Joaquin Valley		Klamath Basin	
Date	**Crop coefficient**	**Date**	**Crop coefficient**
Feb. 3	0.43	May 9	0.08
Feb. 10	0.37	May 14	0.08
Feb. 17	0.41	May 21	0.08
Feb. 24	0.51	May 31	0.08
Mar. 3	0.56	June 7	0.17
Mar. 10	0.60	June 14	0.38
Mar. 17	0.76	June 21	0.61
Mar. 24	0.98	June 30	0.91
Mar. 31	1.12	July 7	1.08
Apr. 7	1.14	July 14	1.14
Apr. 14	1.16	July 21	1.15
Apr. 21	1.21	July 31	1.15
Apr. 28	1.24	Aug. 7	1.15
May 5	1.15	Aug. 14	1.15
May 12	1.09	Aug. 21	1.13
May 19	0.90	Aug. 31	0.9
May 26	0.82	Sept. 7	0.69
June 2	0.58	Sept. 14	0.48
June 9	0.44		

Sources: Estimated by B. Sanden, UCCE Farm Advisor, Kern County (southern San Joaquin Valley); estimated by Harry Carlson, UCCE Farm Advisor (retired), Modoc County (Klamath Basin).

Processing Tomatoes

Crop coefficients for processing tomatoes ranged from 0.15 to 0.20 for the initial growth stage, 1.02 to 1.09 for the midseason stage, and 0.55 to 0.78 for the end of the late-season stage (Hanson and May 2006a). Figure B-2 shows the relationship between canopy cover and crop coefficient developed on the west side of the San Joaquin Valley, while figure B-8 shows crop coefficient over time for different planting dates (Hanson and May 2006b).

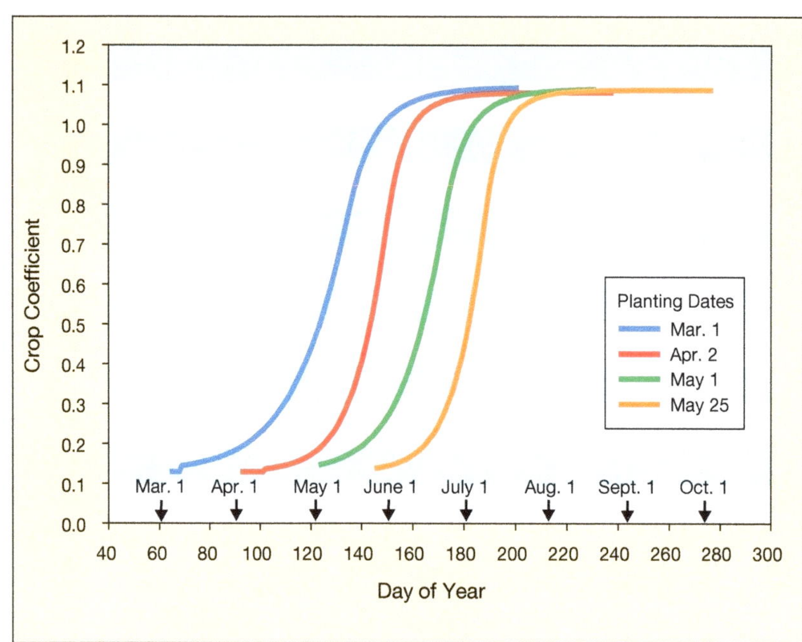

Figure B-8. Crop coefficients of processing tomatoes over time for various planting dates. *Source:* Hanson and May 2006b.

Metric Conversions

English	Conversion factor for English to metric	Conversion factor for metric to English	Metric
inch (in)	2.54	0.394	centimeter (cm)
foot (ft)	0.3048	3.28	meter (m)
acre (ac)	0.4047	2.47	hectare (ha)
acre-inch (ac-in)	102.8	0.00973	cubic meter (m^3)
square foot (ft^2)	0.0929	10.764	square meter (m^2)
gallon (gal)	3.785	0.26	liter (l)
ounce (oz)	28.35	0.035	gram (g)
pound (lb)	0.454	2.205	kilogram (kg)
pound per acre (lb/ac)	1.12	0.89	kilogram per hectare (kg/ha)
pound per acre-inch (lb/ac-in)	44.2	0.22	kilograms per hectare-meter (kg/ha-m)
pound per square inch (psi)	6.89	0.145	kilopascal (kPa)

References

Addink, J. J., J. Keller, C. H. Pair, R. E. Sneed, and J. W. Wolfe. 1980. Design and operation of sprinkler systems. In M. E. Jensen, ed., Design and operation of farm irrigation systems. St. Joseph, MI: The American Society of Agricultural Engineers.

Allen, R., L. S. Pereira, D. Raes, and M. Smith. 1998. Crop evapotranspiration: Guidelines for computing crop water requirements. FAO Irrigation and Drainage Paper 56. United Nations, Rome.

Brennan, D. 2008. Factors affecting the economic benefits of sprinkler uniformity and their implications for irrigation water use. Irrigation Science 26: 109–119.

California Department of Water Resources. 2005. California Water Plan, Update 2005. DWR Web site, http://www.waterplan.water.ca.gov/previous/cwpu2005/index.cfm.

Cavero, J., L. Jiménez, M. Puig, J. M. Faci, and A. Martínez-Cob. 2008. Maize growth and yield under daytime and nighttime solid-set sprinkler irrigation. Agronomy Journal 100(6): 1573–1579.

Chen, D., and W. W. Wallender. 1984. Economic sprinkler selection, spacing, and orientation. Transactions of the American Society of Agricultural Engineers 27(3): 737–743

Cuenca, R. H. 1989. Irrigation system design: An engineering approach. New Jersey: Prentice Hall.

Diamond Plastics Corporation. 2005. PVC pipe for the 21st century. DPC Web site, http://www.dpcpipe.com/.

Frost, K. R., and H. C. Schwalen. 1955. Sprinkler evaporation losses. Agricultural Engineering (August).

Grattan, S. R., W. Bowers, A. Dong, R. L. Snyder, J. J. Carroll, and W. George. 1998. New crop coefficients estimate water use of vegetable, row crops. California Agriculture 52(1): 16–21.

Hanson, B. R. 1995. Practical potential irrigation efficiencies. In Water resources engineering. Proceedings of the First International American Society of Civil Engineers Conference, August 14–18, San Antonio, TX.

Hanson, B. R. 2000. Irrigation pumping plants. Oakland: University of California Division of Agriculture and Natural Resources Publication 3377.

Hanson, B. R., and K. Kaita. 1999. Historical reference crop ET reliable for irrigation scheduling during summer. California Agriculture 53(4): 32–36.

Hanson, B. R., and D. M. May. 2006a. Crop evapotranspiration of processing tomato in the San Joaquin Valley of California, USA. Irrigation Science 24: 211–221.

———. 2006b. New crop coefficients developed for high-yield processing tomatoes. California Agriculture 60(2): 95–99.

Hanson, B. R., and W. W. Wallender. 1986. Bidirectional uniformity of water applied by continuous-move sprinkler machines. Transactions of the American Society of Agricultural Engineers 29(4): 1047–1053.

Hanson, B. R., S. Orloff, and B. Sanden. 2007. Monitoring soil moisture for irrigation water management. Oakland: University of California Division of Agriculture and Natural Resources Publication 21635.

Heermann, D. F., and R. A. Kohl. 1980. Fluid dynamics of sprinkler systems. In M. E. Jensen, ed., Design and operation of farm irrigation systems. St. Joseph, MI: American Society of Agricultural Engineers.

Hunter Industries, Inc. 2009. The handbook of technical irrigation information. Hunter Industries Web site, http://www.hunterindustries.com/Resources/pdfs/Technical/Domestic/LIT194w.pdf.

Iowa State University MidWest Plan Services. 1999. Sprinkler irrigation systems. Ames, IA: MidWest Plan Services Publication MWSP-30. MWPS Web site, http://www.mwps.org/index.cfm?fuseaction=c_Products.viewProduct&catID=720&productID=6479&skunumber=MWPS-30%20%26%20NCR-148&crow=9.

The Irrigation Association. 2000. Center pivot design. Fairfax, VA: The Irrigation Association. Irrigation Association Web site, http://irrigation.org/IAWEB/Core/Orders/category.aspx?catid=2.

Irrigation Training Program. 2008. Texas A&M AgriLife EM-103. South Texas edition. College Station, TX: Texas Water Resources Institute, Texas A&M University.

James, L. G. 1988. Principles of farm irrigation systems. New York: Wiley.

Kasapligil, D. 1990. Row crop sprinkler distribution uniformity in field conditions. Report to the Westlands Water District, CA.

Keller, J., and R. D. Bliesner. 1990. Sprinkle and trickle irrigation. New York: Van Nostrand Reinhold.

López-Urrea, R., F. Martín de Santa Olalla, A. Montoro, and P. López-Fuster. 2009. Single and dual crop coefficients and water requirements for onion (Allium cepa L.). Agricultural Water Management 96: 1031–1036.

Louie, M. J., and J. S. Selker. 2000. Sprinkler head maintenance effects on water application uniformity. Journal of Irrigation and Drainage Engineering 126(3): 142–148.

Montazar, A., M. Sadeghi. 2008. Effects of applied water and sprinkler irrigation uniformity on alfalfa growth and hay yield. Agricultural Water Management 95: 1279–1287.

Natural Resources Conservation Service. 1983. National engineering handbook. In Sprinkle irrigation. Section 15, Chapter 11. Washington, D.C.: USDA, NRCS.

Nderitu, S. M., and D. J. Hills. 1993. Sprinkler uniformity as affected by riser characteristics. Transactions of the American Society of Agricultural Engineers 9(6): 515–521.

Playán, E., R. Salvador, J. M. Faci, N. Zapata, A. Martínez-Cob, and I. Sánchez. 2005. Day and night wind drift and evaporation losses in sprinkler solid-set and moving laterals. Agricultural Water Management 76: 139–159.

Rain Bird Corporation. 1971. Sprinkler irrigation handbook. Glendora, CA: Rain Bird Sprinkler Mfg. Corp.

Sanden, B., J. Mitchell, L. Wu, and S. Leung. 1999. Effects of irrigation nonuniformity on nitrogen and water use efficiency in shallow-rooted vegetable cropping systems (fourth year progress report). Proceedings of the 7th Annual Fertilizer Research and Education Program Conference, November 30, Modesto, CA.

Seginer, I. 1969. Wind variation and sprinkler water distribution. Journal of Irrigation and Drainage Division, American Society of Civil Engineers IR2: 261274.

Snyder, R. L., B. Lanini, D. A. Shaw, and W. O. Pruitt. 1989a. Using reference evapotranspiration (ET_0) and crop coefficients to estimate crop evapotranspiration (ET_c) for agronomic crops, grasses, and vegetable crops. Oakland: University of California Division of Agriculture and Natural Resources Publication 21427.

———. 1989b. Using reference evapotranspiration (ET_0) and crop coefficients to estimate crop evapotranspiration (ET_c) for trees and vines. Oakland: University of California Division of Agriculture and Natural Resources Publication 21428.

Stern, J., and E. Bresler. 1983. Nonuniform sprinkler irrigation and crop yield. Irrigation Science 4: 17–29.

Tarjuelo, J. M., J. Montero, P. A. Carrion, F. T. Honrubia, and M. A. Calvo. 1999. Irrigation uniformity with medium size sprinklers, Part II: Influence of wind and other factors on water distribution. Transactions of the American Society of Agricultural Engineers 42(3): 677-689.

Trout, T. J., and L. F. Johnson. 2007. Estimating crop water use from remotely sensed NDVI, crop models, and reference ET. Proceedings of the USCID 4th International Conference on Irrigation and Drainage, October 3–6, Sacramento, CA.

USDA. 1983. Sprinkler irrigation. U.S. Department of Agriculture Soil Conservation Service National Engineering Handbook, section 15, chapter 11.

Yazar, A. 1984. Evaporation and drift losses from sprinkler irrigation systems under various operating conditions. Agricultural Water Management 8: 439–449.

CPSIA information can be obtained
at www.ICGtesting.com
Printed in the USA
JSHW040941140323
38903JS00007B/66